Standard and Super-Resolution
Bioimaging Data Analysis

Current and future titles in the Royal Microscopical Society—John Wiley Series

Standard and Super-Resolution Bioimaging Data Analysis: A Primer

Edited by

Ann Wheeler
Advanced Imaging Resource MRC-IGMM
University of Edinburgh, UK

Ricardo Henriques
MRC Laboratory for Molecular Cell Biology
University College London, UK

Published in association with the Royal
Microscopical Society

Series Editor: Susan Brooks

WILEY

Registered Office(s)
John Wiley & Sons, Inc., 111 River Street, Hoboken, NJ 07030, USA
John Wiley & Sons Ltd, The Atrium, Southern Gate, Chichester, West Sussex, PO19 8SQ, UK

Editorial Office
The Atrium, Southern Gate, Chichester, West Sussex, PO19 8SQ, UK

For details of our global editorial offices, customer services, and more information about Wiley products visit us at www.wiley.com.

Wiley also publishes its books in a variety of electronic formats and by print-on-demand. Some content that appears in standard print versions of this book may not be available in other formats.

Library of Congress Cataloging-in-Publication Data

Names: Wheeler, Ann, 1977– editor. | Henriques, Ricardo, 1980– editor.
Title: Standard and Super-Resolution Bioimaging Data Analysis: A Primer /
    edited by Dr. Ann Wheeler, Dr. Ricardo Henriques.
Description: First edition. | Hoboken, NJ : John Wiley & Sons, 2018. |
    Includes index. |
Identifiers: LCCN 2017018827 (print) | LCCN 2017040983 (ebook) |
    ISBN 9781119096924 (pdf) | ISBN 9781119096931 (epub) | ISBN 9781119096900 (cloth)
Subjects: LCSH: Imaging systems in biology. | Image analysis–Data processing. |
    Diagnostic imaging–Data processing.
Classification: LCC R857.O6 (ebook) | LCC R857.O6 S73 2017 (print) | DDC 616.07/54–dc23
LC record available at https://lccn.loc.gov/2017018827

Cover design by Wiley
Cover image: Courtesy of Ricardo Henriques and Siân Culley at University College London

Set in 10.5/13pt Sabon by SPi Global, Pondicherry, India
Printed and bound in Malaysia by Vivar Printing Sdn Bhd

10 9 8 7 6 5 4 3 2 1

Contents

6 **Co-Localisation and Correlation in Fluorescence Microscopy Data**

Dylan Owen, George Ashdown, Juliette Griffié and Michael Shannon

List of Contributors

George Ashdown, Department of Physics and Randall Division of Cell and Molecular Biophysics, King's College London, UK

Graeme Ball, Dundee Imaging Facility, School of Life Sciences, University of Dundee, UK

Sébastien Besson, Centre for Gene Regulation & Expression and Division of Computational Biology, University of Dundee, UK

Mario De Piano, Division of Cancer Studies, King's College London, UK

Ahmed Fetit, Advanced Imaging Resource, MRC-IGMM, University of Edinburgh, UK
and
School of Science and Engineering, University of Dundee, UK

Juliette Griffié, Department of Physics and Randall Division of Cell and Molecular Biophysics, King's College London, UK

Aliaksandr Halavatyi, European Molecular Biology Laboratory (EMBL), Heidelberg, Germany

Ricardo Henriques, MRC Laboratory for Molecular Cell Biology, University College London, UK

Gareth E. Jones, Randall Division of Cell & Molecular Biophysics, King's College London, UK

Debora Keller, Facility for Imaging by Light Microscopy, Imperial College London, UK

Kota Miura, Nikon Imaging Center, Bioquant, University of Heidelberg, Germany; National Institute of Basic Biology, Okazaki, Japan; Network of European Bioimage Analysts (NEUBIAS)

Nicolas Olivier, Department of Physics and Astronomy, University of Sheffield, UK

Peter O'Toole, Technology Facility, Department of Biology, University of York, UK

Dylan Owen, Department of Physics and Randall Division of Cell and Molecular Biophysics, King's College London, UK

Thomas Pengo, University of Minnesota Informatics Institute, University of Minnesota Twin Cities, USA

Michael Shannon, Department of Physics and Randall Division of Cell and Molecular Biophysics, King's College London, UK

Stefan Terjung, European Molecular Biology Laboratory (EMBL), Heidelberg, Germany

Jean-Yves Tinevez, Institut Pasteur, Photonic BioImaging (UTechS PBI, Imagopole), Paris, France

Sébastien Tosi, Advanced Digital Microscopy Core Facility (ADMCF), Institute for Research in Biomedicine (IRB Barcelona). The Barcelona Institute of Science and Technology, Barcelona, Spain; Network of European Bioimage Analysts (NEUBIAS)

Claire M. Wells, School of Cancer and Pharmaceutical Sciences, King's College London, UK

Ann Wheeler, Advanced Imaging Resource, MRC-IGMM, University of Edinburgh, UK

Foreword

Imaging is now one of the most commonly used techniques in biological research. It is not simply a means of taking a pretty picture; rather advanced microscopy and imaging are now vital biophysical tools underpinning our studies of the most complex biological systems. We have the capability to study cells in real time, with 3D volumes, analysing biophysical interactions, and now moving towards the ability to see and study individual proteins within individual cells within their native environment. Imaging is one of the most useful tools for understanding the fundamental biology of the cell.

This has been made possible through an incredibly rapid period of microscopy development, which has gone hand in hand with the emergence of new fluorescent tags – such as fluorescent proteins – computing power and the latest developments in engineering. Not only has the technology become increasingly versatile and opened up many new possibilities, but leading manufacturers have also made their microscopes increasingly accessible to non-specialists, resulting in an explosion of data.

All of these developments have left us with a wealth of data, but the images themselves will remain just pretty pictures unless they are analysed appropriately. We are now starting to see an equivalent rapid increase in the development of image analysis, but we are still far from realising its full potential. The basics are vital, and anyone using today's microscopes should also be looking at the best approaches for analysing their data.

This book is extremely timely and looks at some of the key aspects of data analysis; it will serve as an excellent point of reference. Chapter 1 examines the basics of image data and processing which is common to most users. Chapters 2 and 3 builds on this and looks at how to quantify both routine 2D through to more complex 3D image data sets. However, one of the greatest challenges remains the ability to segment our images. To our own eyes, this can be quite obvious, but it still remains a real computational challenge. Segmenting images and image data e.g. in Chapter 3 will be a continuing area of development that will also help us to correlate data with greater precision in the future.

Beyond the images, the microscope is a powerful biophysical tool. We can see and analyse the co-localisation of particles such as proteins, but great care is needed in their analyses as outlined by Dylan Owen e.g. in Chapter 6. We can now go beyond this and start to see interactions that occur over a 5 nm range. This enables the studies of particle–particle interactions such as protein–protein heterodimerisation by using FRET, and although the imaging of these phenomena is relative simple the complexity of the quantification is discussed in Chapter 4.

Not only can the microscope study these natural interactions, but we can also use the light, often with lasers, to manipulate cells and trigger critical events that would otherwise occur in a random fashion making their studies very difficult. The interpretation and controls for FRAP and other photo perturbation methods then needs to be carefully considered (Chapter 5).

Many of the above studies can be undertaken using both fixed and live cell imaging. Whole live-cell imaging and tracking brings its own analytical challenges (Chapter 7) as cells move through three dimensions, pass over one another often changing shape and nature. This needs many of the above elements to come together to help work in such complex sample types.

At the other extreme, from whole cells, the biggest advancement in light microscopy has come from the ability to image below the diffraction limit. Super-resolution microscopy (SRM) is possible through many different strategies. The analysis and interpretation is an area that is often under-appreciated and which can result in misinterpretations. For anyone wanting to undertake SRM, it is essential to understand the limitations, controls and best approaches to their analysis (Chapter 8).

Many of the new microscopical techniques now produce very large data sets. This is especially true for 3D live-cell imaging and SRM. This has developed its own problem, with data analyses now often taking considerably longer than the imaging time itself. The time needed for data analysis has become the most costly element in complex image study. The staff time itself often outweighs the cost of the instrument and consumables and this is why we need to look for automation when handling and analysing the data (Chapter 9), and naturally, once all of this data has been analysed, it is vital to not only present the data in the correct manner, but also to ensure that it is correctly documented and stored (Chapter 10). Only then, can any one image be properly exploited and deliver the required impact.

Peter O'Toole
York
July 2017

1

Digital Microscopy: Nature to Numbers

Ann Wheeler

Advanced Imaging Resource, MRC-IGMM, University of Edinburgh, UK

Bioimage analysis is the science of converting biomedical images into powerful data. As well as providing a visual representation of data in a study, images can be mined and used in themselves as an experimental resource. With careful sample preparation and precise control of the equipment used to capture images, it is possible to acquire reproducible data that can be used to quantitatively describe a biological system, for example through the analyses of relative protein or epitope expression (Figure 1.1). Using emerging methods this can be extrapolated out over hundreds and thousands of samples for high content image based screening or focused in, using emerging technologies, to data at the nanoscale. Fluorescence microscopy is used to specifically mark and discriminate individual molecular species such as proteins or different cellular, intra-cellular or tissue specific components. Through acquiring individual images capturing each tagged molecular species in separate channels it is possible to determine relative changes in the abundance, structure and – in live imaging – the kinetics of biological processes. In the example below (Figure 1.1), labelling of F-actin, a cytoskeletal protein, using a

Standard and Super-Resolution Bioimaging Data Analysis: A Primer, First Edition.
Edited by Ann Wheeler and Ricardo Henriques.
© 2018 John Wiley & Sons Ltd. Published 2018 by John Wiley & Sons Ltd.

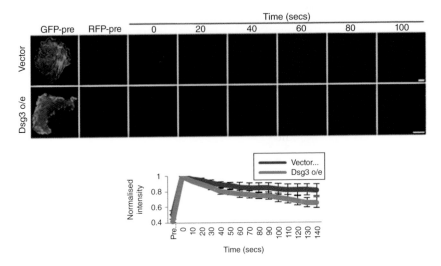

Figure 1.1 Bioimage quantification to determine the dynamics of actin using photoconversion. Tsang, Wheeler and Wan Experimental Cell Research, vol. 318, no. 18, 01.11.2012, p. 2269–83.

fluorescent protein allows measurement of how fast it turns over in moving cells normally, and in a condition where a putative regulator of cell migration DSG3 is overexpressed. It shows that overexpressing DSG3 destabilises actin and causes it to turn over faster. Quantifying the expression and localisation of F-actin in several cells over time it is possible to see how much F-actin it turns over in the course of the experiment, where this happens, and the difference in rate between the two (Figure 1.1, graph). This type of scientific insight into the spatial and temporal properties of proteins is only possible using bioimage analysis and illustrates its use in current biomedical research applications.

In this book we are primarily going to consider quantification of images acquired from fluorescence microscopy methods. In fluorescence microscopy, images are acquired by sensors such as scientific cameras or photomultiplier tubes. These generate data as two-dimensional arrays comprising spatial information in the x and y domain (Figure 1.2); separate images are required for the z spatial domain – known as a z stack – which can then be overlaid to generate a 3D representative image of the data (Figure 1.2). Image analysis applications such as Imaris, Volocity, Bioimage XD and ImageJ can carry out visualisation, rendering and analysis tasks. The most sensitive detectors for fluorescence and bright-field microscopy record the intensity of the signal emitted by the sample, but no spectral information about the dye (Figure 1.3).

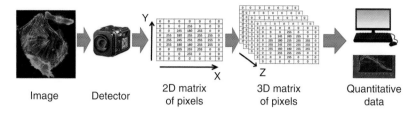

| Image | Detector | 2D matrix of pixels | 3D matrix of pixels | Quantitative data |

Figure 1.2 Workflow for bioimage data capture in 2D and 3D.

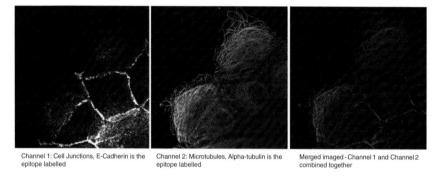

Channel 1: Cell Junctions, E-Cadherin is the epitope labelled   Channel 2: Microtubules, Alpha-tubulin is the epitope labelled   Merged imaged - Channel 1 and Channel 2 combined together

Figure 1.3 Combining channels in fluorescent bioimage analysis. Channel 1 has antibodies raised against E-cadherin labelled with AlexaFluor 568 secondary antibodies. Channel 2 is labelled with primary antibodies raised against Alpha tubulin and secondary antibodies labelled with AlexaFluor 488.

This means effectively that intensity information from only one labelled epitope is recorded. To collect information from a sample which is labelled with multiple fluorescent labels the contrast methods on the imaging platform itself – e.g. fluorescent emission filters, phase or DIC optics – are adjusted to generate images for each labelled epitope, all of which can then be merged (Figure 1.3). Some software will do this automatically for the end user. The final dimension that images can be composed of is time. Taken together, it is possible to see how a 3D multichannel dataset acquired over time can comprise tens of images. If these experiments are carried out over multiple spatial positions – e.g. through the analysis of multiwell plates or tilling of adjacent fields of view – the volume of data generated can considerably scale up, especially when experiments need to be done in replicates. Often the scientific question may well require perturbing several parameters, e.g. adjustment of different hypothesised parameters or structures involved in a known biological process. This means that similar image acquisition and analysis needs to be used to analyse the differences in the biological system.

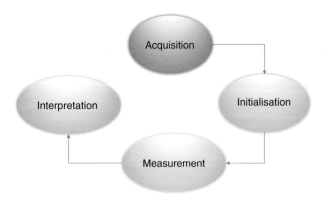

Figure 1.4   The Bioimage analysis workflow.

In these cases although setting up an automated analysis workflow makes sense, to manually quantify each individual image would take a considerable time and would require a substantial level of consistency and concentration. The programming of analysis pipelines does require some work initially but it can be seen as letting the computer automate a large volume of tasks, making the research process more reliable, robust and efficient. Indeed some applications now allow data processing in batches on remote servers, computer clusters or cloud computing.

Biomedical image analysis follows a given workflow: data acquisition, initialisation, measurement and interpretation (Figure 1.4) – which will be discussed in brief in this introductory chapter, followed by a more in-depth analysis in subsequent chapters.

1.1 ACQUISITION

1.1.1 *First Principles: How Can Images Be Quantified?*

Before data can be analysed, it needs to be acquired. Image acquisition methods have been extensively reviewed elsewhere [1, 3, 4]. For quantification, the type and choice of detector which converts incident photons of light into a number matrix is important. Images can be quantified because they are digitised through a detector mounted onto the microscope or imaging device. These detectors can be CCD (charged coupled device), EMCCD (electron multiplying CCD) or sCMOS (scientific CMOS) cameras, or photomultiplier tubes (PMTs). Scientific cameras consist of a fixed array of pixels. Pixels are small silicon semiconductors which use the photoelectric effect to convert

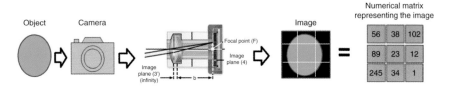

Figure 1.5 How images are digitised.

the photons of light given off from a sample into electrons (Figure 1.5). Camera pixels are precision engineered to yield a finite number of electrons per photon of light. They have a known size and sensitivity, and the camera will have a fixed array of pixels. Photons of light pass from the object to become images through the optical system, until they collide with one part of the doped silicon semiconductor chip or pixel in the camera. This converts the photons of light into electrons which are then counted. The count of 'photo electrons' is then converted into an intensity score, which is communicated to the imaging system's computer and is displayed as an image (Figure 1.5). PMTs operate on similar principles to scientific cameras, but they have an increased sensitivity, allowing for the collection of weaker signals. For this reason they are preferentially mounted on confocal microscopes. Photomultipliers channel photons to a photocathode that releases electrons upon photon impact. These electrons are multiplied by electrodes called metal channel dynodes. At the end of the dynode chain is an anode (collection electrode) which reports the photoelectron flux generated by the photocathode. However, the PMT collects what is effectively only one pixel of data, therefore light from the sample needs to be scanned, using mirrors, onto the PMT to allow a sample area larger than one pixel to be acquired. PMTs have the advantage that they are highly sensitive and, within a certain range, pixel size can be controlled, as the electron flow from the anode can be spatially adjusted; this is useful as the pixel size can be matched to the exact magnification of the system, allowing optimal resolution. PMTs have the disadvantage that acquiring the spatial (x, y and z) coordinates of the sample takes time as it needs to be scanned one pixel at a time. This is particularly disadvantageous in imaging of live samples, since the biological process to be recorded may have occurred by the time the sample has been scanned. Therefore live imaging systems are generally fitted with scientific cameras and systems requiring sensitivity for low light and precision for fixed samples often have PMTs. (https://micro.magnet.fsu.edu/primer/digitalimaging/concepts/photomultipliers.html)

1.1.2 *Representing Images as a Numerical Matrix Using a Scientific Camera*

Although having a pixel array is useful for defining the shape of an object it doesn't define the shading or texture of the object captured on the camera. Cameras use greyscales to determine this. Each pixel has a property defined as 'full well capacity'. This defines how many electrons (originated by photons) an individual pixel can hold. An analogy of this would be having the camera as an array of buckets, which are filled by light. It is only possible to collect as much light as the pixel 'well' (bucket) can hold; this limit is known as saturation point. There can also be too little light for the pixel to respond to the signal, and this is defined as under-exposure.

The camera can read off how 'full' the pixel is by a predetermined number. This is defined as the greyscale. The simplest greyscale would be 1-bit, i.e. 0 or 1. This means that there is either light hitting the pixel or not; however, this is too coarse a measure for bioimage analysis. Pixels record intensity using binary signals, but these are scaled up. Pixels in many devices are delineated into 256 levels, which corresponds to 2^8, which is referred to as 8-bit. The cone of a human eye can only detect around 170–200 light intensities. So a camera, set at 8-bit (detecting 256 levels) produces more information than an eye can compute. Therefore, if images are being taken for visualisation, and not for quantification, then using a camera at 8-bit level is more than adequate. For some basic measurements, 8-bit images are also sufficient (Figure 1.6).

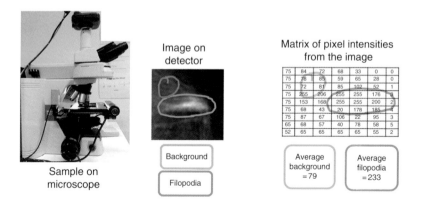

Figure 1.6   Basic quantification of cellular features using 8-bit fluorescent image of F-actin.

It is possible to increase the sensitivity of the pixel further, currently to 12 (4096 or 2^{12}), 14 (16384 or 2^{14}) and 16 (65536 or 2^{16}) grey levels. For detecting subtle differences in shading in a complex sample, the more numerical information and depth of information that can be mined from an image the better the data that can be extracted can be. This also allows better segmentation between noise inherent in the system and signal from the structure of interest (Figure 1.6).

Although this chapter is concerned with bioimage analysis it is essential that the images are acquired at sufficient sensitivity for quantification. Scientific cameras currently can delineate up to 2^{16} grey levels dependent on their specification. The image histogram, is a 1D representation of the pixel intensities detected by the camera. It can be used to determine the distribution of pixel intensities in an image, making it easy to perceive the saturation or under-sampling of an image acquired (Figure 1.7). A saturated signal is when the light intensity is brighter than the pixel can detect and the signal is constantly at the maximum level. This means that differences in the sample can't be detected as they are being recorded at an identical greyscale value, the maximum intensity possible (Figure 1.7). Under-sampling, which means not making use of the full dynamic range of the detector or having information below the detection limit of the detector is not ideal. It means that the intensity information is 'bunched together', and so subtle structures may not be able to be detected (Figure 1.7). Under-sampling is sometimes necessary in bioimaging, for

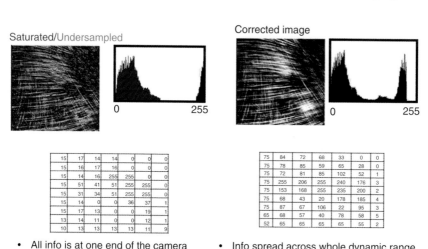

Saturated/Undersampled

Corrected image

0   255

0   255

15	17	14	14	0	0	0
15	16	17	16	0	0	0
15	14	16	255	255	0	0
15	51	41	51	255	255	0
15	31	34	51	255	255	0
15	14	0	0	36	37	1
15	17	13	0	0	19	1
13	14	11	0	0	12	1
10	13	13	13	13	11	9

75	84	72	68	33	0	0
75	78	85	59	65	28	0
75	72	81	85	102	52	1
75	255	206	255	240	176	3
75	153	168	255	235	200	2
75	68	43	20	178	185	4
75	87	67	106	22	95	3
65	68	57	40	78	58	5
52	65	65	65	65	55	2

- All info is at one end of the camera
- Data is squashed
- Can't tell differences between staining

- Info spread across whole dynamic range
- More sensitivity for specific measurements

Figure 1.7 The effect of saturation and under-sampling on bioimage analysis.

instance if imaging a very fast process or when a very weak signal is being collected from a probe which can be photo-damaged. Provided that sufficient signal can be collected for quantitative analysis this need not be a problem. However, best practice is to have the signal fill the whole dynamic range of the detector.

The first and perhaps most important step in bioimage analysis is that images be acquired and quantified in a reproducible manner. This means:

- using the same piece of equipment, or pieces of equipment that are technically identical
- ensuring equipment is clean
- ensuring samples are as similar as possible and prepared similarly
- using the same parameters to acquire data, e.g. same magnification, same fluorescent labels and very similar sample preparation and mounting.

1.1.3 *Controlling Pixel Size in Cameras*

Pixels in scientific cameras are a predefined size, while in PMTs the scan area can be adjusted so that pixel size can be varied (see Section 1.1 on acquisition). The ideal pixel size matches the Nyquist criteria – that is, half the size of the resolution that the objective permits, providing the pixel is sufficiently sensitive to detect the signal of interest. Camera pixel size can limit resolution as it is difficult to spatially separate two small structures falling in the same pixel unless subpixel localisation methods are used, as discussed in Chapter 8. It is very difficult to spatially separate two small structures falling in the same pixel. If a larger pixel size is required it is possible to have the detector electronically merge pixels together. This is generally done when a 2×2 array of pixels or a 4×4 array is combined into one super-pixel. The advantage of this is that there is a 4 (2×2 bin) or 16 (4×4) fold increase in sensitivity since the 'merged pixels' add together their signals. The trade-off is a loss of spatial sampling as the pixels are merged in space. For studies of morphology, the resolution of the camera is important; pixels (i.e. the units comprising the detection array on the scientific camera) are square, and for any curved phenomena the finer the array acquiring it, the better will be the representation curves of the sample. The loss of spatial detail can be problematic if the structures studied are fine (Figure 1.8). Using brighter dyes – that is those with a higher quantum yield of emitted photons per excited photon – and antifade agents to prevent bleaching can help here.

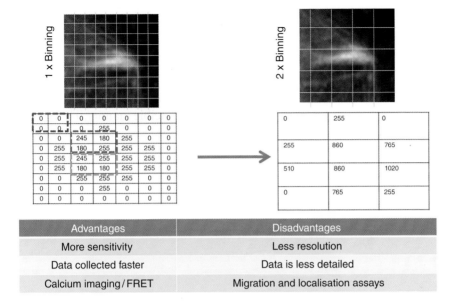

Advantages	Disadvantages
More sensitivity	Less resolution
Data collected faster	Data is less detailed
Calcium imaging / FRET	Migration and localisation assays

Figure 1.8   Binning of pixels to increase speed and sensitivity of Bioimage acquisition.

For studies of protein expression, sensitivity can be important, although the bit depth of the pixel plays a role. If the detector can only detect a fraction of the light being produced because it either meets its saturation point or is under-exposed it causes issues. The epitope will be either not detected or under-sampled because the detector is not capable of picking up sufficient signal for quantification (Figure 1.8).

In studies of fast transient reaction (e.g. calcium signalling), fast exposure and frame rate can be more important than spatial resolution (Figure 1.8). Here, binning can be extremely useful since the sensitivity to an individual pixel may not be sufficient to detect subtle changes in signal. Binning also allows the camera to record data and transfer this electronic information to the computer faster since there are fewer pixels (Figure 1.9).

Detectors have a finite capacity for signal and a certain output speed, and this can be analogised to an array of buckets that have a certain capacity for water and tip it out at a certain rate (Figure 1.10). Knowing the speed of the camera to write the detected information to the computer's disk is important. In live experiments, cameras can detect signals faster than the speed with which the computer can write information to the disk. This is known as a clocking problem and is troublesome because data is collected, but it isn't recorded to the computer disk (Figure 1.9).

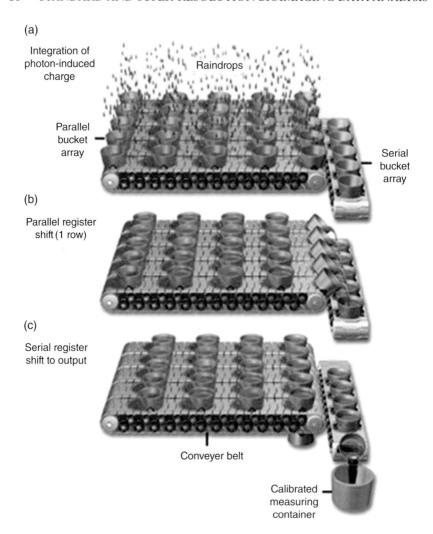

(a)

Integration of
photon-induced
charge

Raindrops

Parallel
bucket
array

Serial
bucket
array

(b)

Parallel register
shift (1 row)

(c)

Serial register
shift to output

Conveyer belt

Calibrated
measuring
container

Figure 1.9 Bucket brigade CCD analogy (Courtesy of Molecular Expressions, Florida state Univeristy, USA, https://micro.magnet.fsu.edu/primer/index.html).

56	76	64	78	34
72	63	89	76	97
82	47	102	83	83
68	83	79	72	65
54	66	52	64	55

Median = 79

Figure 1.10 A 3×3 median filter kernel. The filter size is indicated in orange. This filter smooths the image and denoises it.

The most recent advance in camera technology, sCMOS cameras, can be beneficial because they combine a small pixel size with high sensitivity and fast read time (clocking). They have applications in a wide variety of biological questions where the phenomena to be imaged are small and either transient or entail rapid kinetics. These devices can also be implemented for scanning of large areas in techniques such as light-sheet microscopy due to their large field of view and high-speed acquisition.

Camera manufacturers producing instruments that are suitable for quantitative imaging:

1. Andor Technologies http://www.andor.com/
2. Hammamatsu http://www.hamamatsu.com/
3. Leica Microsystems http://www.leica-microsystems.com/home/
4. Lumenara https://www.lumenera.com/
5. Nikon Instruments https://www.nikoninstruments.com/
6. Olympus http://www.olympus-lifescience.com/en/
7. PCO Instruments https://www.pco-tech.com/
8. Photometrics http://www.photometrics.com/
9. QImaging http://www.qimaging.com/
10. Motic Instruments http://www.motic.com/As_Microsope_cameras/
11. Zeiss Microscopy http://www.zeiss.com/microscopy/en_de/software-cameras.html

1.2 INITIALISATION

Initialisation is the step where bioimages are prepared for quantification. In most cases, the image generated by the system will not be immediately suitable for automatic quantification, and most analysis requires the computer to have a set of very similar artefact-free images for the analysis algorithms to function correctly. It is thus critical to minimise image features that may corrupt or hamper the analysis framework to be used. The dominant aberrations in the detection system are caused at three levels: (a) the sample itself, (b) the microscope or scanner's optical properties through which the image is formed and (c) the detector. These aberrations need to either be minimised or removed entirely so that the signal to be processed in the image is clearly distinguished from the noise which is otherwise present in the sample. Techniques used to do this such as filtering, deconvolution and background subtraction, and registration in x, y, z and colour channels needs to be carried out.

1.2.1 The Sample

The sample to be imaged may contain artefacts or structures that are challenging to image, which makes it difficult to acquire good images for analysis. The key to good analysis is excellent sample preparation. Dyes and antibodies need to be optimised so that they are bright enough to be within the linear range of the detector. Ideally the background from non-specific binding or antibodies or other probes would be reduced. The fixation and processing of samples would be optimised. Even with these strategies in place, a digital camera can only acquire a 2D image of a biological structure which is itself 3D. This means that out of focus light from around the focal plane is present in the image, which may obscure the signal from in-focus light. Confocal systems minimise out-of-focus light in acquired images by physical methods involving the use of pinholes. However, since most light in a sample is out of focus, only a small fraction of light is allowed through the pinhole increases the need for bright labelling [1]. Further inappropriate fixation or storage can damage samples, and sample mounting is also challenging because 3D samples can be squashed or shrunk. For studies in thick tissue, where the sample will be cut into a sequence of individual thin slices that will be imaged, there can be issues with collating these images back into a virtual 3D representation of the tissue [2].

1.2.2 Pre-Processing

Not all parts of images may need to be processed, and the regions to be measured may need to be turned into separate images. The imaging system may acquire data in a format that is not compatible with the analysis algorithm. Some imaging applications store images in individual folders (Leica LAS, Micromanager) and data may need to be moved to an analysis server. Due to the nature of image acquisition rescaling, techniques such as histogram equalisation may be necessary. All of these steps contribute to the pre-processing. Most applications enable this and would have some kind of image duplication function or a means of saving the pre-processed data separately from the raw data. The raw image data must be retained to comply with scientific quality assurance procedures which are discussed in Chapter 10, which deals with presentation and documentation.

1.2.3 Denoising

Denoising is removal or reduction of noise inherent in the sample and imaging system which masks the signal of interest. Cameras and PMTs are not perfect, and are subject to several sources of noise. Noise is defined as electrons that are read by the camera that have not been generated by photons from a sample, for example,

- *Shot noise:* This is caised by random electrons generated by vibration inside the camera or PMT.
- *Dark current:* PMTs and cameras have a baseline number of electrons that it reads even when there is no light. Manufacturers will usually set this to be a non-zero value, and PMTs in particular have a base current from photocathode to anode even in the absence of light. Measuring the dark current on a system is useful, because if this value falls below the normal value, it helps the end user determine that there is a problem with the camera. A low dark current can be achieved by cooling the detector; often CCD and EMCCD cameras are cooled for this reason.
- *Read noise:* The photoelectric silicon semiconductor has a range of accuracy, e.g. although it will usually generate two electrons per photon sometimes it may generate one and sometimes three. The accuracy of the read noise depends on the quality of the pixel chip. The number of electrons yielded per photon can be described as the quantum yield.
- *Spectral effects:* Neither PMTs nor cameras produce a linear number of photoelectrons per incident photon across the visible spectrum. At 500 nm, a camera may produce four electrons per photon and at 600 nm it may produce three and at 700 nm, just one. If correlations are being made between two different dyes or fluorophores, it is important to take into consideration what the 'spectral performance' of the detector is.
- *Fixed pattern noise:* Some cameras have random noise caused by spurious changes in charge across the pixel array. Other types, sCMOS in particular, suffer from fixed patter noise, which means that, due to manufacturing or properties of the camera itself, certain parts of the camera have a higher noise level than others. This is often in a fixed pattern, although it can consist of individual 'hot' (i.e very noisy) pixels. This noise pattern can be subtracted from an image.

All scientific cameras and PMTs from reputable manufacturers will include a table and datasheet describing the performance of their instruments. This can be useful to study at the outset of an experimental series where Bioimage analysis is to be done.

1.2.4 *Filtering Images*

Noise is inherent in all bioimages; this may be introduced because of shortcomings with the detector as described above. This type of noise is described as non-structural background, and is low-frequency, and constant in all images. Another source of noise is introduced because the detector can only acquire images in 2D while biological samples are 3D, so out-of-focus light, or issues with labelling the sample may cause the desired signal to be masked. This type of noise is high frequency and can have structural elements. One of the most frequently used methods for initialising images for bioimage analysis is filtering. By using a series of filters it becomes possible to remove most of the noise and background, improving the signal-to-noise ratio. This is generally achieved by mathematical operations called deconvolutions.

In a nutshell, this involves deconvolving the numerical matrix that makes up the bioimage with another number array; they can contain different numbers depending on the desired effect on these images. The technical term for these arrays is kernels, and denoising involves filtering images using kernels.

Detector noise and non-homogenous background from the sample can be removed by a process called flat fielding. This is acquiring an image with a blank slide at the settings used to acquire the bioimages, and subtracting this background noise image from the data. Some image analysis programs can generate a pseudo flat field image if one has not been acquired. This method can be very effective with low signal data if the noise is caused by the detector. 'Salt and pepper' noise can be evened out by using a median filter. A median filter runs through each pixel's signal, replacing the original pixel signal value entry with the median of its neighbours. The pattern of neighbours is called the "window" (Figure 1.10).

The effect is nonlinear smoothing of the signal, but edges of the images suffer as the median value of the edge will involve a null value, which means that a few edge pixels are sacrificed when using this method. Often images generated from PMTs suffer from this type of noise because of shot noise and read noise on the detectors. Other types of filters that can reduce noise in samples are as shown in Figure 1.11a:

- Smooth filter: A pixel is replaced with the average of itself and its neighbours within the specified radius. This is also known as a mean or blurring filter.
- Sigma filter: The filter smooths an image by taking an average over the neighbouring pixels, within a range defined by the standard deviation of the pixel values within the neighbourhood of the kernel.

(a)

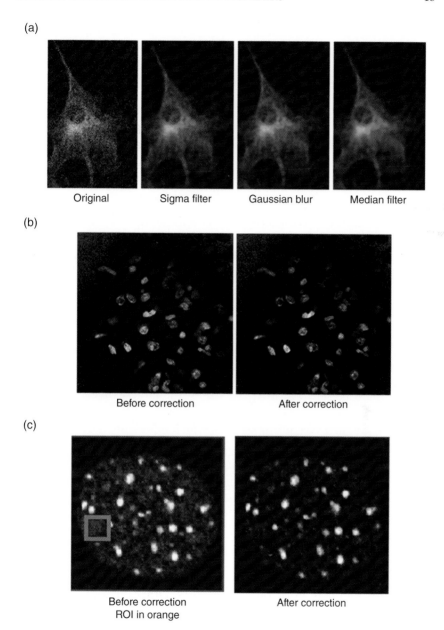

(b)

(c)

Figure 1.11 Initialisation using filtering (a) Illustrative example of image filtering taken from the Image J webpage https://www.fiji.sc, (b) Example of rolling ball background subtraction: left-hand side is before correction, and right-hand side after, (c) Using ROI subtraction.

- Gaussian filter: This is similar to the smoothing filter but it replaces the pixel value with a value proportional to a normal distribution of its neighbours. This is a commonly used mathematical representation of the effect of the microscope on a point of light.

In epifluorescence images there is often a vignette of intensity across the image. This is a result of the illumination in these systems where a mercury halide or LED illuminator is focused into the centre of the field of view to be imaged, provided it is correctly aligned. The bulb will not give an even intensity of illumination; rather the illumination follows a Gaussian distribution. In well-aligned microscopes this means that the image is brightest in the centre and dimmer at the edges. If there is a problem with the alignment of the illuminator, there can be an intensity cast across the image where potentially one of the corners or part of the image is brighter than another. To remove this issue, in ImageJ a 'rolling ball' background correction algorithm designed by Castle and Keller (Mental Health Research Institute, University of Michigan) is implemented (Figure 1.11b). Here a local background value is determined for every pixel by averaging over a very large kernel around the pixel. This value is hereafter subtracted from the original image, hopefully removing large spatial variations of the background intensities. The radius should be set to at least the size of the largest object that is not part of the background [3].

In better-aligned systems or systems which inherently have more even illumination such as confocals, noisy background can be caused by other effects. For instance, uneven illumination caused by "scan lines" in confocal transmitted light images can be removed using the native FFT bandpass function present in ImageJ and other software packages. When detector noise or bleaching is an issue, this can be accounted for by measuring the mean intensity of the region in an image where there is known background and then subtracting the mean value of this region. Although this reduces the net intensity value in an image, it can emphasise relevant data (Figure 1.11c). Removing the high frequency noise caused by labelling, light interference in the sample can be more challenging. Different types of filters can assist with this, and this subject is discussed at greater length in Chapter 3.

1.2.5 Deconvolution

Deconvolution is a method which is used to remove out-of-focus light completely from an image. It is based on the premise that an image is a convolution of the imaged sample with the system used to image it – in the case of light microscopy, the sample and the microscope. No system is optically perfect and objective lenses are a primary cause of aberrations in an image. They suffer from multiple aberrations, predominantly spherical and chromatic, and have artefacts in flatness of field. High-quality objectives such as Plan-Apochromat are corrected for all of these across the visible spectrum but are more expensive than most other objectives. In particular, aberrations in the axial dimension can be particularly problematic for light microscopes. Any lens may do a fairly reasonable job of focusing light in 2D, but 3D focus is more challenging, and lenses tend on average to perform half as accurately in the third (axial) dimension as in x and y. Abbe's law summarises the resolution of a light microscopy image.

$$d = \frac{\lambda}{2n \sin \theta}$$

where d is the resolution of the system, λ is the wavelength of emitted light, $n \sin \theta$ is the numeric aperture of the objective (the numeric aperture is the half angle that light can propagate through the objective). Only the most simplistic of imaging system consists of just an objective. Fluorescent systems will also have dichroic mirrors and filters as well as other moving parts. All of these will slightly distort or absorb photons on their path to the detector. Each distortion, though incremental, adds up. Experts in optics tend to combine this source of error in the optical system in one metric: the point spread function (PSF). This describes how an optical system images an infinitely small and perfectly spherical point of light (Figure 1.12). Inevitably, the refraction of light due to imperfection in the optical system will mean that the point of light is distorted.

Since all images consist of many points of light, knowing about how a given imaging system distorts one point of light means that it's possible to extrapolate this onto an image and 'deconvolve' out the real signal from the distortions introduced by the imaging system [4]. Most bio-image processing applications provide some type of deconvolution, although specialist packages such as Autoquant or Huygens specialise in this. Many depend on an artificial point spread function which will be generated based on the wavelength of emitted light and the magnification and numerical aperture of the objective used. Naturally, contributions from dirty lenses or misaligned optics are not taken into account when generating an artificial PSF. For heavily used instruments

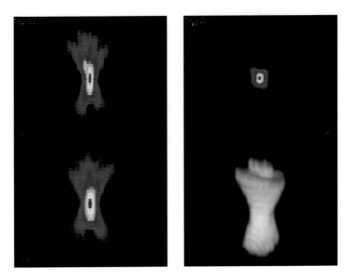

Figure 1.12 Experimental point spread functions: By Howard Vindin (own work) [CC BY-SA 4.0 (http://creativecommons.org/licenses/by-sa/4.0)], via Wikimedia Commons.

it may be advisable to generate an imaged PSF, for example through the observation of small-scale fluorescent beads (e.g. 100 nm Tetraspeck beads – Thermo Scientific). Several algorithms are provided for deconvolution, the most commonly used being:

- *Nearest neighbour approaches:* These are fundamentally two-dimensional, they are classified for the purposes of this discussion as *deblurring algorithms*. As a class, these algorithms apply an operation plane by plane to each two-dimensional plane of a three-dimensional image stack. For example, the nearest-neighbour algorithm operates on the plane z by blurring the neighbouring planes ($z + 1$ and $z - 1$, using a digital blurring filter), then subtracting the blurred planes from the z plane. This has the disadvantage that it can reduce the signal-to-noise ratio, and it can add in sharpened points caused by overlapping signal in the z domain in places where it doesn't belong. The advantage of this approach is that it's computationally light, at the expense of degrading the signal and introduction of artefacts.
- *Iterative deconvolution:* This is an image restoration approach where out-of-focus light is either discarded or brought into focus. The algorithm works by iterating through a set of parameters that best represent the 'in focus' image. To start with, an estimate of the object is performed; generally this is the raw image. This estimate is convolved

with the PSF, and the resulting 'blurred estimate' is compared to the original input image. A metric (figure of merit) of restoration indicates the improvement in the images compared to the original. This metric is then used to adjust the estimated parameters to reduce the error criterion in the subsequent iteration. This process is repeated until the error criterion is minimised or reaches a specified threshold. The final image is the object estimate at the last iteration. Richardson–Lucy deconvolution is a popular implementation of this method [5, 6]. These iterative algorithms often use the likelihood error criteria defined in probability theory. Maximum likelihood estimation (MLE) is a method of *estimating* the parameters of a statistical model given certain observations, e.g. the blurred object image, by finding parameters that maximise the likelihood of making the observations given the parameters. Several commercial applications use these approaches: Huygens, Imaris, Nikon Elements, Carl Zeiss and ImageProPremier [7].

Iterative statistical algorithms are more computationally intensive when compared to non-iterative methods and can take significantly longer to reach a solution. However, they may restore images to a slightly higher degree of resolution than filtering. These algorithms also have the advantage that they impose constraints on the expected noise statistic (in effect, a Poisson or a Gaussian distribution). As a result, statistical algorithms have a more subtle noise policy than simply regularisation, and they may produce better results on noisy images. However, the choice of an appropriate noise statistic may depend on the imaging condition, and some commercial software packages are more flexible than others in this regard. Processing on a server for deconvolution can be a major advantage for a large batch of deconvolution datasets, as they will often take up a significant amount of processing capability of a desktop machine. Further information about deconvolution can be found here:
https://micro.magnet.fsu.edu/primer/digitalimaging/deconvolution/deconvolutionhome.html

1.2.6 *Registration and Calibration*

Spatial calibration of images: Pixels all have a defined size so the amount of physical space of images on a detector will depend on the optical elements (e.g. the microscope lens) that are projecting light onto it. If a small ruler, called a graticule, is imaged it is possible to determine, for a given optical system (e.g. a tissue culture microscope with 10× lens), what the actual size detected by each pixel is, that is, if 100 μm on the graticule is visualised across 50 pixels then one pixel will measure across 2 μm. Many commercial systems will come pre-calibrated, and this information will be included in the image header file, but home-built systems may not be pre-calibrated, and systems where the detector or objective lens has been changed for experimental purposes may also not be correctly spatially calibrated. If you have a high-precision stage, it becomes possible to calculate the pixel size calibration by moving an object for a well-defined distance in the stage. The observed distance travelled in images can then be correlated to the distance set on the stage to get the pixel calibration. For measurement of spatial parameters such as size and shape of imaged objects it is essential to include these measures.

Image registration is technically described process of 'overlaying two or more images of the same scene taken at different times, from different viewpoints, and/or by different instrument settings (e.g. different fluorescent channels)' [8]. Registration geometrically aligns two or more images, which can comprise different channels or different x and y or z planes. Registration algorithms may need to be applied in the following cases:

- Datasets where more than one fluorescent channel is used. Chromatic aberrations will mean that the colours may focus to slightly different places on a camera or detector.
- Datasets collected over time, where the structure may move or the system may drift.
- Datasets collected by multiview analysis, e.g. tiling or mosaicing of an image.

Each of these cases is very different, and the algorithms programmed to address these differ; however, they do work on similar workflows whereby features are detected and matched, the geometric functions required to map or transform the image are created, and the geometric transform is applied. Registration of many biological images can be challenging due to the structures involved and the need to align spatial positions and different channels; image processing experts have generated several tools for this [9]. Libraries, plugins and other tools for image registration and stitching are supplied in software packages such as Amira, Arivis and Vaa3D for this. In some cases the software may ask for features or fiducial markers which can be used to remap the images [10]. In other cases autocorrelation routines or propagation methods can be used, particularly to address system drift. Several open-source plugins for registration are available and can be installed into

ImageJ/Fiji (http://imagej.net/Category:Registration). Often these will have been developed with specific use cases in mind: e.g. subpixel registration of super-resolution microscopy images; stitching and multiview reconstruction of large tissue or cell areas; registration of light sheet microscopy data. Some examples of registration algorithms are: TrakEM2 [11], SURF+affine transformation [12] UnwarpJ, [13] and V3D, [14] and BrainAligner [15].

1.3 MEASUREMENT

Once the noise has been removed from the image, generating useful numerical data for measuring samples can be started. The pixel array from an imaging device, CCD, EMCCD or sCMOS camera or PMT generates a numerical matrix. Hence bioimage measurement is, simply put, sampling this matrix and performing mathematical operations yielding numeric descriptors of the data. Once the image is corrected for any inherent aberrations and is pre-processed so that it is as close a representation of the original object imaged as possible, it is time to move forwards and gather numerical data.

For these features to be measured they must be segmented (i.e. identified as interesting areas to analyse, e.g. nuclei are segmented for imaging when nuclei counting occurs) and then quantified. This topic is very complex, which is why Chapters 2 and 3 of this book are dedicated to the subject, but it lies at the heart of bioimage analysis. Typical features in a light microscopy image to be measured would be: **Size and shape** of biological structures, as a non-exhaustive list: nuclei, endosomes, cytoskeletal components in cells. In tissues it might be neurons, blood vessels or populations of stem cells. **Location** where certain epitopes are, in respect to reference organelles, is often asked. One of the simpler use cases is whether a given protein is localised in the nucleus or in the cytoplasm, although this can be extrapolated to many cases, in particular in developing organisms. **Motion** or **kinetics** of structures could be cells crawling, delivery of cargo on a microtubule or looking at cells developing their fate. **Concentration**, the amount of a given epitope in a specific location in a cell or tissue, is a frequently asked research question. This can be used for examples applied to fluorescence recovery, after photobleaching (FRAP), which would look at how much protein turns over in a given place. Concentration studies are also useful when investigating the role of individual components of a biological complex to its function, e.g. if protein x is removed does the complex still form or only partially form?

These parameters in image analysis would generate the following types of numerical output:

Measuring shape or location would generate Cartesian coordinates (x,y,z) and metrics describing area, length width or the perimeter of a structure, e.g. a nucleus (Figure 1.13). This is done by selecting a tool which puts a contour over the image and the number of pixels it intersects with – or for area are inside a closed contour – are counted. With area the number of pixels intersected by the line and the number of pixels inside this area are also counted.

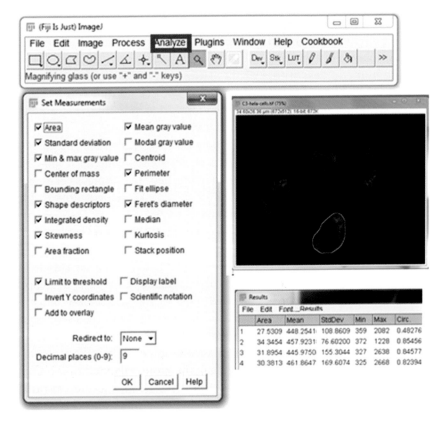

Figure 1.13 Using ImageJ to select parameters of shape and intensity in an image of nuclei (blue). Here nuclei have been manually segmented using a contour – yellow line. Measurements have been set in ImageJ, and the numerical results output. The area of each nucleus, mean, standard deviation, maximal and minimal intensity are computed. The circularity (Circ) of the nuclei are also computed.

Measuring intensity: Here instead of the number of pixels being counted, the intensity value of each of the pixels is counted. The mean, maximum, minimum and standard deviation of these values can easily be determined. It is also possible to sum the values up to give an integrated sum for this (Figure 1.13).

Measuring kinetics: This is a convolution of measurement of either shape or intensity with the time domain.

There is a very wide variety of both commercial and open-source software which can perform measurements on bioimages. A list which covers most of the frequently used applications is given in Table 1.1.

Once the measurement parameters have been developed they can be stored and applied to large data samples. Specific software has been written to better support these high content or big data applications; these are discussed in the Chapter 9.

1.4 INTERPRETATION

Once the most interesting and relevant metrics for a given question have been identified statistical analysis can be applied to this. This can be done through packages such as SPSS, R, Graphpad, Prism, MATLAB, Minitab or by using bespoke analyses. There will be different statistical methods for analysing the spatial distribution of clusters or determining whether one or more features in an image correlates or co-locates with another, and the relevant tests will depend on the question being asked. These types of analysis are described in Chapter 6 and 8. Bioimage quantification analysis has historically relied overmuch on the t-tests, and while they are useful in comparing two populations that vary with one another, they are often not the best method of statistically interpreting bioimage data. Differences in shape or signal intensity between multiple samples can be analysed using analysis of variance (ANOVA) tests which show if there is a significance change in the population sampled. To determine which individual sample in the group analysed by ANOVA is different, post hoc testing of one or more metrics, e.g. the area of a nucleus or the elongation of an axon, can be carried out using Bonferroni or Dunnett's test. Bonferroni cross-correlates each group with the other groups to see which ones are significantly different from the others (many to many analysis). Dunnett's test allows a selected group to be compared with all of the others (one to many analysis). The Tukey HSD test is also useful for identifying arithmetic means of a group which are

Table 1.1 List software for measuring and processing bioimages.

Image Analysis package	Open-source or manufacturer	Data types analysed or function	Reference link
Acapella	Perkin-Elmer	High content screens of 2D images, large volumes	http://www.perkinelmer.co.uk/category/image-analysis-software
Amira + Avizo3D	FEI	Visualisation and quantification of 2D and 3D data, electron microscopy data in particular	https://www.fei.com/software/avizo3d/
Arivis Vision 4D	Arivis	Modular software for working with multichannel 2D, 3D and 4D images of almost unlimited size independent of available RAM	https://www.arivis.com/en/imaging-science/arivis-vision4d
BioimageXD	Open-source – Universities of Jyväskylä and Turku in Finland, MPI-CBG, Germany	3D quantification and segmentation of fluorescent images.	http://www.Bioimagexd.net/
Bisque	Open-source – UCSB, USA	Store, visualise, organise and analyse images in the cloud	http://Bioimage.ucsb.edu/bisque
BigDataViewer	Open-source – MPI-CBG, Dresden Germany	BigDataViewer is a reslicing and quantifying browser for terabyte-sized 3D multiview image sequences	http://imagej.net/BigDataViewer
Cellprofiler + Cellprofiler Analyst	Open-source – Broad Institute USA, [16]	Quantitatively measure phenotypes from many 2D images automatically and machine learning	http://cellprofiler.org/

Name	Source	Description	URL
Cell cognition	Open-source – IMBA Vienna	Development of a fast and cross-platform image analysis framework for fluorescence time-lapse microscopy in the field of bioimage informatics.	http://www.cellcognition.org/
Definiens	Definiens	Accelerates development of quantitative tissue-based image analysis. using machine learning	http://www.definiens.com/
Fiji/ImageJ/ ImageJ 2	Open-source – Worldwide but led by Wayne Rasband (Ex NIH, USA) and Curtis Rueden, UW-Madison LOCI, USA) [16]	Flexible Java image processing program with a strong, established user base and thousands of plugins and macros for performing a wide variety of tasks	http://imagej.net/Fiji/ Downloads
Icy	Open-source Institute Pasteur Paris	Software resources to visualise, annotate and quantify bioimage data, using a friendly user interface and coded in Python and Java.	http://icy.Bioimageanalysis.org/
Imaris	Bitplane, Oxford Instruments	3D and 4D real-time interactive data visualisation, quantification and management	http://www.bitplane.com/ imaris/
Ilastik	Open-source – EMBL Heidelberg Germany and HHMI Janelia Farm USA	*ilastik* is a simple, user-friendly tool for interactive image classification, segmentation and analysis	http://ilastik.org/

(Continued)

Table 1.1 (Continued)

Image Analysis package	Open-source or manufacturer	Data types analysed or function	Reference link
ImageProPremier + Autoquant	MediaCybernetics	2D image analysis software to capture, process, count, classify, measure and share and also enables deconvolution	http://www.mediacy.com/imagepropremier
ImageTool	University of Texas USA	Free C++ image processing and analysis program for pathology and fluorescence 2D images for Windows platforms	http://compdent.uthscsa.edu/dig/itdesc.html
Kalaimoscope	Open-source – MPI-CBG Dresden, Germany	Object tracking in 2D and 3D	http://www.kalaimoscope.com/product.html
Leica LAS	Commercial, Leica Microsystems	Processing and quantification of 2D and 3D images acquired on Leica microscopes using Leica acquisition software	http://www.leica-microsystems.com/home/
MetaMorph/MetaExpress	Molecular Devices	GPU accelerated image analysis of 2D, super-resolution and large volume datasets; some 3D image processing supported	https://www.moleculardevices.com/systems/metamorph-research-imaging/metamorph-microscopy-automation-and-image-analysis-software
NIS Elements	Nikon Instruments	Processing and quantification of 2D, 3D and super-resolution images acquired on Nikon microscopes using Nikon acquisition software	https://www.nikoninstruments.com/en_GB/Products

Name	Source	Description	URL
BiG	Open-source – Bioimaging Group, EPFL Switzerland	Development of new algorithms and mathematical tools for advanced processing of biomedical images, e.g. SpotTracker, Deconvolution Lab	http://bigwww.epfl.ch/
TrackMate	Open-source – Institute Pasteur, France	TrackMate is your buddy for your everyday tracking of 2D and 3D objects	http://imagej.net/TrackMate
Vaa3D	Open-source – Allen Institute of Brain Science, USA	Explores big 3D/4D/5D images with giga-voxels, extracts complex surface objects from images and performs comprehensive analyses such as brain connectome mapping	http://www.alleninstitute.org/what-we-do/brain-science/research/open-science-research-tools/vaa3d/
Visiopharm	Visiopharm.	Quantitative digital pathology solution providing flexible, powerful quantification of tissue properties in a fast, objective and reproducible way	http://www.visiopharm.com/
Zen	Zeiss Microscopy	Processing and quantification of 2D and 3D images acquired on Zeiss microscopes using Zeiss acquisition software	http://www.zeiss.com/microscopy/en_de/home.html

significantly different from one another because the means of several different samples can be tested. If the variance between multiple parameters is of interest, particularly in high content screens, principal component analysis can be particularly helpful. When comparing between populations it is important to bear in mind that many statistical tests assume that the data is normally distributed; this is true of t-tests, ANOVA and any post hoc tests associated with ANOVA. This type of test is called a parametric test. However, many bioimaging experiments, e.g. using CrispR knock-in or siRNA knock-out of genes will cause the distribution of phenotypes to become skewed or not normally distributed. This means that parametric statistical tests can no longer be used and non-parametric analysis based on the median or rank of data is more appropriate. Non-parametric tests which are commonly used in bioimage analysis are Mann–Whitney, which is the non-parametric analogue of the t-test or Kruskal–Wallis which is a non-parametric analogue of ANOVA. An in-depth description of statistics is beyond the scope of this book, but some the following sources of information are useful.

- Nature Statistics for Biologists collection. http://www.nature.com/collections/qghhqm
- http://www.wormbook.org/chapters/www_statisticalanalysis/statisticalanalysis.html
- Statistical and Data Handling Skills in Biology, Roland Ennos, Pearson Education Ltd ISBN: 9780273729495
- Statistics for Terrified Biologists, Helmut van Emden, Wiley ISBN: 978-1-4051-4956-3

Statistical analysis tools for data mining and informatics analysis that can be applied to large multivariate datasets, such as high content screening include: 'Analytics' applications such as the open-source Cellprofiler Analyst, Bisque and KNIME (https://www.knime.org/) or commercially available SpotFIRE (http://spotfire.tibco.com/), AcuityXpress from molecular devices and others that have toolboxes to enable multiparametric analysis, data visualisation and pipelines into comparative analysis with genomic, transcriptomic and proteomic analysis datasets. It is more routine, for smaller datasets to prepare the data for presentation at this point in the experimental process. Strategies for this are discussed in Chapter 10.

Ann Wheeler would like to thank Dr Ricardo Henriques and the IGMM advanced imaging facility users for constructive comment on this chapter.

1.5 REFERENCES

1. Spector, D. and R. Goldman, 2005. *Basic Methods in Microscopy Protocols and Concepts from Cells: A Laboratory Manual.* s.l.: Cold Spring Harbor Press ISBN 978-087969751-8.
2. Husz, Z.L., N. Burton, B. Hill, N. Milyaev and R.A. Baldock, 2012. *Web tools for large-scale 3D biological images and atlases.*, BMC Bioinformatics., pp. 13:122. doi: 10.1186/1471-2105-13-122.
3. Sternberg, S.R., 1983. *Biomedical Image Processing*, IEEE Computer, pp. Volume: 16, Issue: 1, 22–34, 10.1109/MC.1983.1654163.
4. Swedlow, J.R., 2003. *Quantitative fluorescence microscopy and image deconvolution.* Methods Cell Biol., pp. 72:349–67. PMID:14719340.
5. Richardson, W.H., 1972. *Bayesian-Based Iterative Method of Image Restoration.* Journal of the Optical Society of America, pp. 62 (1): 55–59. doi:10.1364/JOSA.62.000055.
6. Lucy, L.B., 1974. *An iterative technique for the rectification of observed distributions*, Astronomical Journal., pp. 79 (6): 745–754. Bibcode:1974AJ...79...745L. doi:10.1086/111605.
7. Lane, R.G., 1996. *Methods for maximum-likelihood deconvolution.* Journal of the Optical Society of America, pp. vol. 13, Issue 10, pp. 1992–1998 doi. org/10.1364/JOSA.13.001992.
8. Zitova, B. and J. Flusser, 2003. *Image Registration Methods, a Survey.* Image and Vision Computing, pp. 21, 21977–1000.
9. Wang C.-W., Ka S.-M. and A. Chen, 2014, *Robust image registration of biological microscopic images.* Scientific Reports 4, Article number: 6050, p. doi:10.1038/srep06050.
10. Alvaro J.P. and M.B. Arrate, 2015, *Free Form Deformation Based Image Registration Improves Accuracy of Traction Force Microscopy.* Plos One, p. 10.1371/journal.pone.0144184.
11. Cardona A., S. Saalfeld, J. Schindelin, et al., 2012, *TrakEM2 Software for Neural Circuit Reconstruction.*PLoS ONE, p. 10.1371/journal.pone.0038011.
12. Ess, A., T. Tutelaars and L. Gool, 2008, *Speeded Up Robust Features (SURF).* Comput. Vis. Imag. Understand., pp. 110, 346–359.
13. Wang, C. and H. Chen, 2013, *SIFT Improved Image Alignment Method in application to X-ray Images and Biological Images*Bioinfo., pp. 29, 1879–1887.
14. Peng, H., Z. Ruan, F. Long, J.H. Simpson and E.W. Myers, 2010. *V3D enables real-time 3D visualization and quantitative analysis of large-scale biological image data sets.* Nat. Biotechnology, pp. 28, 348–53 doi:10.1038/nbt.1612.
15. Peng, H., P. Chung, F. Long et al., 2011, *BrainAligner: 3D registration atlases of Drosophila brains.* Nat. Methods, pp. 8, 493–498 doi:10.1038/nmeth.1602.
16. Schneider, C.A., W.S. Rasband and K.W. Eliceiri, 2012, *NIH Image to ImageJ: 25 years of image analysis.* Nature Methods, pp. 9, 671–675 doi:10.1038/nmeth.2089.

2

Quantification of Image Data

Jean-Yves Tinevez

Institut Pasteur, Photonic BioImaging (UTechS PBI, Imagopole), Paris, France

2.1 MAKING SENSE OF IMAGES

2.1.1 *The Magritte Pipe*

Vision is a fascinating sense – if we stop taking it for granted. For instance, look at the picture in Figure 2.1a. What do you see?

It is likely that you subvocalised something in the line of 'it's a pipe'. But it's not a pipe; it is a painting of a pipe. More precisely, it is a set of coloured pigments spatially arranged in a specific manner. More accurately, this figure is a portion of a famous painting by René Magritte, called the 'La trahison des images' or 'The treachery of images'. The original painting has a text line at the bottom that states 'Ceci n'est pas une pipe' or 'This is not a pipe'. Yet, your vision immediately, and possibly involuntarily, identified correctly what the painting represents. It goes a step further: you made the identification with the actual object represented – you said, 'It *is* a pipe'. Even more, if René Magritte had added some undulating fumes above the bowl, you could have concluded

(a)   (b)

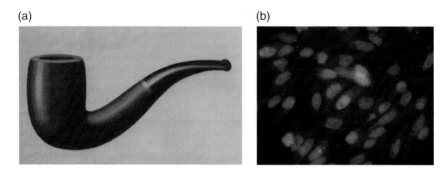

Figure 2.1   (a) Magritte, "The treachery of Images" extract, Los Angeles County Museum of Art. (b) Cells stained in immune-fluorescence. Reproduced with kind permission of Design and Artists Copyright Society, UK.

that the pipe contained burning tobacco, ready to smoke. But these fumes would have been other coloured pigments. Do you see how quickly we could make sense of them? We could almost instantaneously state that there was one object in this painting, that this object was a pipe, standing upright, mouthpiece facing right, and that there was no tobacco lit in it.

The ability of our brain to straightforwardly interpret images (among other representations) has fascinated artists and scientists alike for more than a century [1]. Image analysis would be its scientific counterpart. It is about making sense of images, thanks to computers, and is both an area of research and a paramount tool in science. As imaging technologies and devices bloomed, extracting scientific information from images became a central preoccupation in science and engineering. Quantification of image data can be seen as a subset of image analysis techniques that provides answers in the shape of numbers, extracted from images. We turn to this tool when the questions we ask are about objects: how many objects can be found in an image, what is their size, their shape, how are they oriented, where are they respective to another type of object, how are they spatially arranged, etc. We start with raw images, and extract objects and analyse them further to yield quantitative information from them.

Emulating human vision with computers is the subject of a research field called 'computer vision'. We are more interested in biomedical image analysis, where we work mainly with images acquired from microscopes or other imaging devices specific to life sciences. Though they focus on different challenges, the work done in these two fields percolated both ways.

2.1.2 *Quantification of Image Data Via Computers*

What can you see on Figure 2.1b?

This time, it is likely that you silently exclaimed 'a picture of fluorescently labelled cells!' But we have to go one step further in our raw analysis. If Magritte's painting was made of coloured pigments, this image is simply a list of integer numbers. Biosciences now live in the digital era, and it is likely that all analyses will be done via the computer, therefore dealing with numbers at their lowest level. Breaking image analysis into its simple constituents, our task is to identify objects from a list of integers, and make measurements on them. For instance, in Figure 2.1b, we might be interested in knowing the rate of multinucleated cells, which requires identifying the nuclei and the cells separately, then counting them.

Compared to the power and immediateness of our vision, computer image analysis is surprisingly difficult. You could identify, locate and orient the pipe in a few milliseconds, without any a priori or context information. Doing the same with a computer is almost an impossible task. Why should we rely on computers then? Picture yourself as the reviewer of an article and compare these two (fictitious) sentences:

> From the image displayed in Figure 2.X, it is obvious that bacteria are found close to colonic crypts.
>
> Analysis of colon images reveals that more than 90% of all the bacteria are found within 10 µm of a colonic crypt entry (Figure 2.X, $N_b = 19856$ bacteria, $N_c = 568$ crypts). A comparison with numerical simulations where bacteria attach randomly on the colon surface reveals that the observed spatial distribution is not random.

They both say the same thing, but the second one relies on quantification to make a scientific statement. The field we work in has now become very competitive. Over the years, reviewers have developed a natural resilience for articles containing blunt assessments such as "it is obvious that". They feel that the authors are trying to intimidate an observation into being a scientific fact. Journals with a significant impact factor will require you to support your statements with proper quantification. Journals don't trust the sole researcher's eyes to quantify images, because a brain controls these eyes, which is nowadays under massive pressure to get published. Computers, however, are fully objective. They don't even care about themselves, and can only produce answers in the form of numbers. There is little trifling that can be done with numbers, and a quantified observation is objective. If your work involves imaging, then image analysis has to become one of the tools you master.

The human eye can be used for measurements, however. It is for instance perfectly possible for you to count cells in an image, provided they can be seen. Often, this can be good enough, and is termed 'manual quantification' in publications. Still, quantification via image analysis allows you to get much more from your images. For instance, suppose you are interested in the impact of two groups of drugs on the nucleus morphology. Using your own eyes you may be able to determine that the first group makes the nuclei 'round', and the second 'elongated', and make your point. Quantification enables you to say that the *roundness* (which we will define later) of nuclei is above 0.9 for the first group and below 0.6 for the second. Again, the two approaches carry the same scientific conclusion. However, with numbers, you might be able to pick a third group of drugs significantly different from the others for which the roundness is between 0.6 and 0.8. Numbers, the fruit of quantification, are the key to a finesse and level of detail that are otherwise unmatched. It is impossible not to quote Lord Kelvin on this matter:

> When you can measure what you are speaking about, and express it in numbers, you know something about it; but when you cannot express it in numbers, your knowledge is of a meagre and unsatisfactory kind; it may be the beginning of knowledge, but you have scarcely in your thoughts advanced to the state of science. [2]

Computers also excel in dealing with the automation of repetitive tasks. Though trivial, this challenge is increasingly common in life science. Indeed, the subtleties of the processes we study often involve repeating an experiment numerous times and analysing tens to thousands or more images in the same way. The self-esteem we have leads us legitimately to spend time programming or using a computer to repeat a single analysis task on these images, rather than performing the analysis ourselves, one image after the other.

We can guess from these initial musing that doing quantification involves mastering computers and being familiar with some mathematical concepts. Though these areas might be distant from our initial background and interest, this should not be frightening or discouraging. Quite the contrary; image analysis is a fantastic, thrilling and exciting field. Its benefits largely outweigh the efforts it involves, and the pleasure provided by building an analysis script that performs successfully is its own reward.

2.2 QUANTIFIABLE INFORMATION

Quantification is about extracting numbers from images, and it is our final goal. To reach it, we need to detect and segment objects, which is discussed in the next chapter. Once we have these objects, we can interrogate them for interesting numerical features. We quickly list in this section some of the typical features in life science that researchers are interested in measuring.

2.2.1 *Measuring and Comparing Intensities*

Fluorescence microscopy coupled to immunofluorescence labelling of fluorescent protein expression gives a great way to directly measure the abundance of a target protein in a sample. Since fluorescence microscopy operates in a linear regime, it directly yields pixel intensities that are proportional to the amount of fluorophore at the object location. Still, the proper acquisition of images to support this quantification can be surprisingly difficult to achieve. James Pawley listed 39 factors [3] that, when acquiring fluorescence images, could compromise the validity of subsequent analysis of a single image. Acquiring a batch of images depicting several conditions for comparison is even harder, since you have to ensure that variance in results does indeed come from the sample differences rather than from the instability of the acquisition system. Beyond the acquisition step, measuring intensity on a valid digital image is about segmenting objects using a first channel then computing (for instance) the mean pixel value for each object in a second channel. It is desirable to have two separate channels for segmentation and quantification. Indeed, many segmentation techniques rely on the pixel values to operate. If you segment objects using the signal coming from the channel you want to quantify, you might involuntarily detect only the brightest objects or have a segmentation accuracy that varies with the intensity you are trying to quantify. This will bring a bias in your intensity quantification. This is compounded in 3D analysis where aberrations present in the microscope make quantification even more challenging.

We rely on segmentation rather than detection algorithms, and all software packages for bioimage processing (Table 3.1) ship the tools for intensity measurements. A segmentation result yields the list of the pixels that compose an object, from which different quantities can be calculated (Tables 2.1 and 2.2). Mean intensity is well suited if you are just interested in an average amount inside objects, for instance to compare protein expression. But since you have a whole distribution of intensities,

Table 2.1 Standard object intensity descriptors.

Intensity analysis descriptor	Function
Mean, median	Gives information on the main intensity value of an object – useful to compare the brightness of several objects
Sum	Computes the total fluorescence content inside an object – sensitive to small changes in the segmentation
Standard deviation, variance	Gives information about the dispersion of the intensities – useful to check whether an object is homogenous or not
Skewness	Tells whether intensities are distributed symmetrically (skewness 0) or mainly around low (positive skewness) or large (negative skewness) values – for inhomogeneous objects, can tell what intensities dominate
Kurtosis	Tells whether a symmetric distribution is spiky (positive kurtosis) or broad (negative kurtosis) compared to a Gaussian distribution (kurtosis 0)

you can use any descriptive statistics quantity. The standard deviation or variance will tell about the dispersion of intensities inside single objects and might be relevant if you are interested in the inhomogeneity of staining. The skewness and kurtosis values will tell you about the shape of the intensity distribution and give insights into whether some values are preferred over others in objects that are otherwise of the same mean. Suppose that you follow cells stained for a soluble protein that aggregates in punctate structures upon stimulus (Figure 2.2, [4]). The total protein quantity is conserved, so the mean intensity in a cell will stay the same when stimulated. But when aggregation occurs, the intensity standard variation will increase. The time course of an intensity-based value allows you to track a biological process at the single-cell level.

2.2.2 Quantifying Shape

Achieving proper shape quantification is often the main reason to go to segmentation algorithms, which are discussed in Chapter 3. Ultimately, the most accurate shape descriptor for an object is the list of coordinates that compose its contour. This profusion of information is accurate but useless per se, and we have to condense it to a few discrete quantities that we can relate to more easily. Some of these quantities commonly used in bioimage analysis are listed in Table 2.2, and reviewed below.

Table 2.2 Summary of some object shape descriptors in 2D and 3D.

Object segmented		Definition	Application
2D	3D		
Area	Volume	Total area or volume enclosed in the contour	Quantify the size of objects
Perimeter	Surface	Length or surface of the contour	Quantify the size of the interface between objects and the exterior
Circularity, roundness[1]	Sphericity	2D: $4\pi A/P^2$ if A and P are the area and perimeter of the object. 3D: ratio of the surface of a perfect sphere of identical volume over the surface of the object	First step in quantifying the 'regularity' of objects – a value close to 1 indicates that the object resembles a circle or a sphere; for very elongated objects, values approach 0
Ellipse fit, aspect ratio	Ellipsoid fit, prolate, oblate, scalene	Fit an ellipse or an ellipsoid on the contour. Aspect ratio value, prolate, oblate or scalene shape is determined from the principal axes of the ellipse or ellipsoid	When the fit is valid, this is useful to tell whether an object is elongated or not (comparing the lengths of principal axes). In 3D, objects can be classified as prolate object (rugby ball shape), oblate object (coin shape) or scalene object (nothing special) based on the principal axes length of the fit. When objects have a clear elongation, the fit also yields their orientation, which is great to quantify the direction objects point at
Solidity	Solidity	Area (2D) or volume (3D) ratio between the object and its convex hull	Quantifies how rough the object is: does it have crevices, indentations and irregularities? Ranges from 0 (very irregular shapes) to 1 (convex shapes)

[1] Roundness and circularity are sometimes used interchangeably. In ImageJ, however, the roundness is the inverse of the aspect ratio, defined for the ellipse fit.

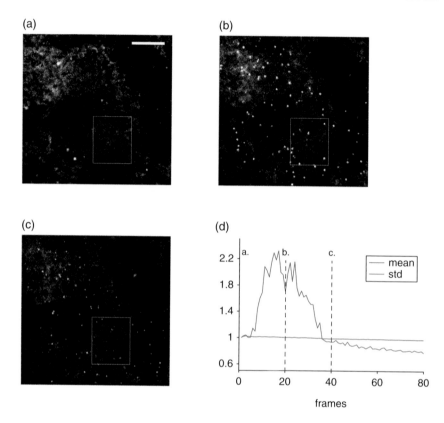

Figure 2.2 (a–c) GFP-NEMO–expressing fibroblasts. Upon treatment with IL-1, the soluble protein aggregates in small punctate, transient structures that can be detected by live-cell fluorescence microscopy, and disappear 10 minutes after stimulation. Scale bar: 10 μm. Panels a, b and c show the first, 20th and 40th frame respectively of the movie post-stimulation. (d) Mean intensity and its standard deviation calculated inside the yellow ROI in preceding panels, normalized with their value at frame 1. Sample courtesy of Nadine Tarantino and Emmanuel Laplantine, Institut Pasteur, Paris, France.

2.2.2.1 Object Size Descriptors

The size of an object is often the initial shape descriptor to go to. It is given by the area for 2D objects and the volume for 3D objects. This is at the heart of many image analysis questions, looking as a comparator at the size of objects can give several useful biological insights. It is important for the segmentation technique to be robust when measuring sizes. If you rely on thresholding pixel intensity for instance, a small variation in the value of the threshold may generate a change in where the object contour is set. It is therefore essential to have a robust method for segmenting objects, as discussed in the next chapter.

For object images, a larger threshold will eat some part of the object away, while a low threshold will inflate it. Even if the contour moves by only one or two pixels, this small change may cause a large variation of the object size value. Images where objects are very sharp are less affected by this problem and are therefore desirable for size quantification. It is advisable when setting out on automated analysis to set up a small 'training set' of images, which are quantified manually, and then compare this to the automatic parameters. You can then probe the robustness of your method by adjusting manually the threshold around the set value by a small amount, and checking if the changes induced are within acceptable bounds.

2.2.2.2 Contour Size Descriptors, Circularity and Roundness

The size of the contour itself is given by the perimeter in 2D, and the surface in 3D. It is useful when you need to quantify the amount of accessible interface between an object and the exterior volume. For instance, microvilli are organelles that tend to increase the interface surface for a given fixed cell volume. This large surface is important for e.g. secretion and absorption, and the quantification of surface helps evaluating the potency of the cell to perform these processes.

If you fix the volume of a 3D object, vary its shape and measure the resulting surface, you will find that the minimal surface is achieved for a perfect sphere. For a 2D object, the minimal perimeter is achieved by a circle. Every deviation from these two shapes will increase the surface or perimeter. Large surfaces are produced by highly irregular shapes, such as those of microvilli. This can be used to produce a numerical feature that will reflect quantitatively the regularity of objects: the sphericity (in 3D) or the circularity (in 2D). They are defined for a single 3D object as the surface of an ideal sphere that would have the same volume that of the object, divided by its actual surface. For 2D objects, it is the ratio of the square of the perimeter of an ideal circle with the same area as that of the object, divided by the square of its actual perimeter. Or equivalently:

$$C_{2D} = \frac{4\pi A}{P^2}$$

with A and P being the area and perimeter of the object. With this definition, circular objects have a circularity of 1. This values goes towards 0 as the object interface becomes irregular. For instance, the circularity is 0.79 for a square, 0.61 for an equilateral triangle, 0.26 for a rectangle of sides 10 and 1, and roughly 0.22 for a five-pointed star. A problem you might encounter is that the exact definition varies from software to software. Sometimes the circularity is not the ratio *squared*, and sometimes it is the inverse, ranging from 1 to infinity, that is reported. Sometimes the word 'roundness' is used, but it might be a definition that relies on the fit by an ellipse, described below. In all cases, carefully study and cite the documentation attached to the software when reporting shape quantities. The software listed in Table 3.1 may rely on different definitions for the same quantity, giving apparently different results with the same datasets.

2.2.2.3 *Measuring Elongation and Orientation*

Sphericity and circularity are really about how large is the contour of an object with respect to its size. The circularity is insufficient, for instance, to discriminate between an elongated rectangle with sides 1 and 10, and five-pointed star. The elongation of an object is captured by another set of quantities derived from fitting an ellipsoid (3D) or an ellipse (2D) to its contour. The ellipse or ellipsoid might actually very poorly fit the contour, but it is adequate to derive interesting quantities. The aspect ratio is the length of the major axis of the ellipse divided by the length of its minor axis. It is 1 for circular objects, and increases for elongated objects. For the rectangle with sides 1 and 10, the aspect ratio is roughly 10. For the star, it is around 1. In 3D, the ellipsoid fit yields three principal axes, and the shape of the objects can be classified further depending on its elongation. Scalene objects have no particular elongation and are characterised by their three principal axes having values that are close. Some objects resemble a cigar or a rugby ball (cf. American football), and are said to be prolate. They have one of their principal axes significantly larger than the other two. Oblate objects look flat like a disc or a coin. One of their principal axes is significantly smaller than the other two.

Ellipse or ellipsoid fits can already tell a lot about object shape, at a rather idealised level, discarding any regularity information. Another advantage is that as soon as an object has an elongated shape, the fit gives access to the direction of this elongation. The major axis angle gives the orientation of a 2D object in a plane. This can be valuable information when trying to assess whether all objects are aligned, and if they are pointing to a clear common direction. This is particularly important when investigating developmental biology questions such as tissue morphogenesis or fate specification as the alignment or misalignment of cells and tissues can be interpreted to have significant biological meaning. In live analysis this can be extrapolated onto the analysis of the vector of movement and can be useful in looking at cell homing in chemotaxis or endocytic movement.

2.2.2.4 Quantifying Smoothness

So far we have reviewed quantities that tell us first whether our objects are big or small, then if their shapes are round or elongated. The next variables in this progression towards shape complexity quantify their ruggedness. Are their contours smooth or spiky? To do so we need to rely on a transformation of the contour called the convex hull.

Imagine that we are holding a five-branched star flat against a table. Its shape can be described as a polygon with ten points, five for the branches tip, and five close to the star centre. Imagine now that we stretch a rubber band around this star. If left to relax, the rubber band will not stick to the star contour. It will not bend to reach the interior points, but will join the five branches' tips, taking the shape of a regular pentagon. The shape that this rubber band takes up is called the convex hull of the star. Its mathematical definition is a bit more involved. If we consider the set of all points that build a contour, the convex hull is the subset of these points that builds a shape containing all the lines that join any possible two points on the original contour.

The convex hull has an interesting property for us: the convexity means that its contour does not have any crevices, so it gives a rather smooth contour for a possibly rough object. The solidity shape descriptor exploits it to quantify the ruggedness of an object and tells whether it is smooth or rough. It is defined as the area ratio (or volume ratio in 3D) between the object and its convex hull. For smooth objects, the convex hull lies on the object contour, and the solidity is equal to 1. For very rough objects, the convex hull is much larger in size than the object, and the solidity approaches 0.

2.2.3 Spatial Arrangement of Objects

The quantities we defined in the previous two sections characterised the intensity of single objects and their shape, and required object contours. We now turn to the quantification of how objects are arranged with respect to each other, relying on variables belonging to the spatial statistics field. Interestingly, this domain found its first uses in ecology [5] and epidemiology [6], where objects are typically animals, plants and trees or diagnosed cases. Initially, their location was extracted via non-imaging techniques such as direct measurements on the field. More recently, satellite imaging and aerial photography also made good use of the detection techniques described above. At the opposite side of sizes, microscopy data makes advantageous use of these quantities to decipher interactions between proteins or organelles [7].

As we will see, spatial statistics are defined over a set of objects, and just need their location to proceed, so you can simply rely on object detection techniques to compute them. We will present just two quantities that characterise the spatial arrangement of a pool objects. There are a vast number of other quantities in the field that build up in complexity, but these two are enough to get started for common biological imaging applications.

2.2.3.1 Dispersion and Aggregation Index

The simplest quantity that describes a spatial arrangement of objects is the dispersion. It is simply a measure of how close objects are in a set. Several definitions exist, based on distance covariance. Another approach uses the distance to the nearest neighbour, averaged over all objects [8]. It offers the obvious direct measure of the distance from one object to its closest neighbour, with dispersion assessed by the standard deviation. This can already reveal or discard possible interactions between objects. Interestingly, we can compute the same distance from an object of one type to the closest object of another type. This is very useful for object-based co-localisation studies, which are the subject of Chapter 6.

Another use of this quantity is to tell whether objects are aggregated or dispersed, by comparing their measured value to what it is for a random spatial distribution. Let's suppose that we are studying numerous small organelles that can be detected in a cell. Let ρ be the density of N objects in the cell, defined in 3D by N/V with V the volume of the cell (assessed, for example, by confocal microscopy) or N/A in 2D with A the cell area. If the N organelles were spread in the cell randomly, the distance to the nearest neighbour would be on average:

- In 2D: $r_u = \dfrac{1}{2\sqrt{\rho}}$ [8]

- In 3D: $r_u = \left(\dfrac{3}{4\pi\rho}\right)^{1/3} \Gamma(4/3) \simeq 0.554 \dfrac{1}{\rho^{1/3}}$ [9]

The Clark–Evans aggregation index R is the ratio of the mean nearest-neighbour distance measured by the above value. If R is smaller than 1, it means that your objects are aggregated, they have a shorter distance to their neighbour than it would be if their location was random in the cell. They might have interactions or external forces that drive them closer to each other, or they might simply be segregated within the cell. It also means that parts of the cell are devoid or have scarcer objects than where the aggregate is. If R is greater than 1, then your objects are dispersed, as if they were repulsing each other. The largest value of R is bounded, and approaches 2.15 in 2D for a uniform distribution of objects on a hexagonal lattice. So as the aggregation index increases from 0 to its maximum value, the spatial pattern it describes moves from aggregated to random, then uniform.

In practice, you have to turn to statistical data analysis software packages, such as R [10], to compute the aggregation index. They add tests for statistical significance, and correction for border effects.

2.2.3.2 Clustering

The aggregation index can be confused for distributions that resemble a uniform distribution. For instance, Figure 2.3a depicts 2000 particles randomly spread over a $200 \times 200\,\mu m^2$ area. To simulate aggregation, the particles that are roughly $30\,\mu m$ away from the centre have been displaced towards the centre by at most $8\,\mu m$. This creates a small aggregate near the centre, then some dispersion just outside the aggregate. Away from the centre, the particles were not moved, so their distribution is still random. The Clark–Evans aggregation index cannot distinguish this: its value is 1.007.

There are slightly more sophisticated tools, such as the Ripley's K function that can deal with situations like this. They probe the spatial repartitions of objects at several scales, and can therefore deal with aggregation at some scale and dispersion at another scale. This function is just about telling how many objects there are within a sphere of radius r around an object, for every r, averaged over all objects. Its mathematical definition is:

$$K(r) = \frac{1}{N\rho} \sum_i \sum_{j \neq i} \delta\left(d_{ij} < r\right)$$

where $\delta\,(d_{ij} < r)$ is equal to 1 if the distance from object i to object j is smaller than r, and 0 otherwise. Note that it is a function, not just an index. It can probe several scales of the spatial distribution.

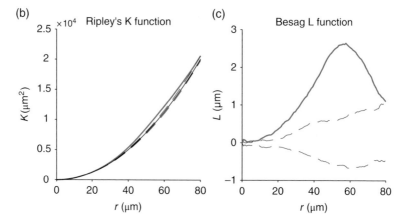

Figure 2.3   (a) Simulation of 2000 particles spread over 200×200 µm. The particules were spread uniformly, then their position was updated by a quantity equal to $dr = r^{-0.05r}$ towards the center with r being their distance to the center. In effect, this moves mainly the particles in a radius of 25 µm towards the center, creating a small aggregation below this radius and depletion at this radius. Beyond 50 µm away from the center, the particle positions are left unchanged, and their distribution remains random. (b) The Ripley's K function for this particle distribution (blue) and for a similar random distribution (black). Dashed red lines: 95% confidence interval for complete spatial randomness assessed by numerical simulations. (c) The Besag L function for this distribution. The function peaks around 60 µm, giving a broad estimate of the aggregate size.

Let us consider the case where we have objects spread randomly. Then:

$$\sum_{j \neq i} \delta \left(d_{ij} < r \right) = N \frac{\pi r^2}{A}$$

in 2D and

$$\sum_{j \neq i} \delta \left(d_{ij} < r \right) = N \frac{4/3 \pi r^3}{V}$$

in 3D. So $K_u(r) = \pi r^2$ in 2D and $K_u(r) = 4/3\pi r^3$ in 3D for a uniform distribution. We now proceed as for the aggregation index, and compare how our measured K compares to the theoretical one obtained for uniform distribution. If for some r the experimental curve is below the uniform distribution curve, this means that for a neighbourhood of radius r there are fewer objects than for a uniform distribution. Therefore at this scale, we might conclude that objects are dispersed. Conversely, if the experimental curve is above the uniform distribution curve, it means that objects are clumped in an aggregate sensibly of this size. In practice, actual calculations of the K function include corrections for border effects. Indeed, the spheres centred on a particle that is near the edge of the region of interest will be truncated. This will introduce a strong bias and to avoid it, our simple mathematical definition needs to take this into account.

The Figure 2.3b is a plot of the K function for the simulated particles. The presence of an aggregate is picked up by the K function, as it is significantly above the K_u. However, the K function is not immediately easy to grasp, as it diverges proportionally with r^2. The Besag L function offers a better representation of the same information. It is defined as

$$L(r) = \sqrt{\frac{L(r)}{\pi}} - r$$

When the objects are randomly spread, the L function is 0 everywhere. Clustering manifests when the curve becomes positive for inspected sizes, and dispersion manifests when the curve becomes negative. Numerical simulations can offer a confidence interval, by generating numerous K and L curves for random distribution of objects from which we extract the 95% confidence interval. In Figure 2.3c, though the presence of a cluster is hard to guess with the naked eye, the Besag L function shows clearly that there is one, gives a measure of its size, and shows that the clusterisation is significant. The function can still be improved to yield a more immediate significance test [7, 11]. These topics are discussed in more detail in Chapter 6.

2.3 WRAPPING UP

The topics discussed above lie at the heart of image analysis. An understanding of the basic mathematical principles which they rely on is essential to performing scientifically relevant quantitation. Most of the image analysis tools will have some means of carrying out the calculations and yield this quantifiable information. However, the actual implementation of these calculations can be somewhat variable depending on the software and its intended audience. If in doubt, it is always recommended to seek and read the documentation of the software, to properly understand exactly what it does to your data mathematically.

We restricted ourselves to survey static quantities: they are defined for a single image (2D or 3D). There is a whole world of quantifiable information to be found in time-lapse images, fuelled for instance by single-particle tracking algorithms, optic-flow techniques and all image analysis techniques able to harness time. They are reviewed in Chapters 5 and 7. Also, some information can be extracted from bulk images, without relying on object extraction. This is the case for instance for co-localisation techniques, reviewed in Chapter 6. All of the quantities we describe here rely on some objects to be extracted from images, moving from low-level information (the pixels) to a higher level (meaningful objects). How these objects are extracted is the subject of Chapter 3.

2.4 REFERENCES

1. McCloud, S., 1994. *Understanding comics: [the invisible art]*: HarperPerennial.
2. Kelvin, W.T., 1891. *Popular lectures and addresses. Nature series*, London, New York,: Macmillan and Co.
3. Pawley, J., 2000. *The 39 steps: a cautionary tale of quantitative 3-D fluorescence microscopy*. BioTechniques. **28**(5): p. 884–6, 888.
4. Tarantino, N. et al., 2014. *TNF and IL-1 exhibit distinct ubiquitin requirements for inducing NEMO-IKK supramolecular structures*. J Cell Biol. **204**(2): p. 231–45.
5. Goodall, D.W., 1952. *Quantitative Aspects of Plant Distribution*. Biological Reviews. **27**(2): p. 194–242.
6. Beale, L. et al., *Methodologic Issues and Approaches to Spatial Epidemiology*. Environmental Health Perspectives, 2008. **116**(8): p. 1105–1110.
7. Lagache, T. et al., 2013. *Analysis of the Spatial Organisation of Molecules with Robust Statistics*. PLoS ONE. **8**(12): p. e80914.
8. Clark, P.J. and F.C. Evans, 1954. *Distance to Nearest Neighbor as a Measure of Spatial Relationships in Populations*. Ecology. **35**(4): p. 445–453.
9. Chandrasekhar, S., 1943. *Stochastic Problems in Physics and Astronomy*. Reviews of Modern Physics. **15**(1): p. 1–89.
10. Baddeley, A., *Clark and Evans Test*, in *R*. p. http://www.inside-r.org/packages/cran/spatstat/docs/clarkevans.test.
11. Lagache, T., *Spatial Analysis*, in *Icy*. p. http://icy.bioimageanalysis.org/plugin/Spatial_Analysis.

3

Segmentation in Bioimaging

Jean-Yves Tinevez

Institut Pasteur, Photonic BioImaging (UTechS PBI, Imagopole), Paris, France

3.1 SEGMENTATION AND INFORMATION CONDENSATION

We call 'segmentation' the process of identifying an object in an image and getting its contour. The name is inherited from computer vision, where an image is segmented in higher-level components, or segments. The same word is also used when objects are simply identified and localised in an image, without yielding its contour. To get the rate of multinucleated cells from Figure 2.1b of Chapter 2, we simply need to identify separately cells and nuclei. Getting their contour would be superfluous, unless an accurate count requires it.

Segmentation is very often the first step when doing quantification. It moves you from low-level information (a list of integers) to a higher level (a set of objects, with a proper class, e.g. nucleus or cell). Interestingly, a huge amount of information condensation happens on the way. The raw image that comprises Figure 2.1b takes about 8 MB in memory ($1024 \times 1344 \times 3$ channels $\times 16$ bit detector). A segmentation algorithm

Standard and Super-Resolution Bioimaging Data Analysis: A Primer, First Edition.
Edited by Ann Wheeler and Ricardo Henriques.
© 2018 John Wiley & Sons Ltd. Published 2018 by John Wiley & Sons Ltd.

yields a count of 35 nuclei and 33 cells. So our segmentation result is made of two positive integers, occupying 8 bytes, and 1 million less memory space than the raw data. We discarded most of the information present in the raw data but still only the two numbers above allow for a scientific conclusion ('the rate of multinucleated cells is low'). The raw list of 4 million numbers is not that helpful as is. This notion of information condensation is absolutely critical. The example above is an analysis based on initial segmentation. Other image analysis techniques, such as pixel-based co-localisation, optical flow, etc. are not based on segmentation, yet offer useful information condensation that can lead to a proper scientific conclusion. In general, the quantification of image data is about condensing abundant low-level information of the raw images into a sparse but meaningful set of numbers. This can be generalised further: a lot of processes related to success in life involve condensing a profusion of data to yield a useful conclusion (defending a thesis, writing a good CV, applying for a grant, guessing if your significant other has a new haircut...). Here we focus on the list of integers that comprises an image.

3.1.1 A Priori Knowledge

Here is another example taken from [1]. In Figure 3.1a is a floating cell, imaged in bright-field. Thanks to a pulsed laser, we pierced a tiny hole in the active cortex at its equator. The rest of the actin cortex contracts, and the internal pressure pushes against the bare membrane and generates this bleb that you see growing on the top left side of the cell. For this study, we needed to quantify the maximal size of the bleb and its speed of growth. We knew that bleb and cell shapes could be approximated by two circles, so we derived a methodology that fitted a two-circle shape around the cell (Figure 3.1b). If we follow the information quantity along the analysis process, we see a massive drop in the quantity of information. We first filtered the image, restricting the range of pixel values. The segmentation of the cell generated a contour with roughly 400 couple of integers, which we fitted with a two-circle shape, yielding three real numbers, the two radiuses and the distance between the two circle centres. All other values could be derived from these three. From the raw image where you need to list all the pixel values to the final results where we just have three real numbers, the condensation is enormous. Again, these three numbers are the only ones that are important for us.

It is crucial to note that the information condensation was not conducted blindly. We knew what the typical noise in the image was, what

(a) (b)

(c) (d)

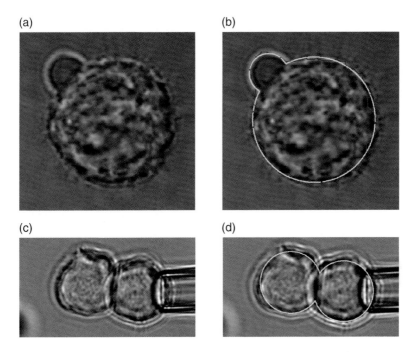

Figure 3.1 (a) L929 cell observed in bright-field. A pulsed 405 nm laser was used to ablate the cell cortex in its periphery, which generated a bleb that can be see growing on the top-left part of the cell. (b) The cell image, overlaid with the result of segmentation (red) and the two-circle fit (yellow). (c) Two cells aspired in a micropipette, observed in bright-field. (d) The cells image overlaid with the two-circle fit (yellow).

were the image features that defined the cell's contour, and the model that could describe the cell shape economically. Another imaging modality would have led to a completely different approach. So you need to have a priori knowledge on your image content: What are the objects in the image? what is their approximate size? How were they imaged? Is this a fluorescence image or a transmitted light image? Is the staining specific to the target objects? Is it an optically sectioned portion of the sample, or the whole 3D volume? The answers to this kind of question will immediately elect some analysis algorithms and discard others.

3.1.2 An Intuitive Approach

The disregard or improper exploitation of a priori knowledge can lead to false scientific claims. Conversely, we can exploit our vision in an intuitive manner to get inspiration for the next step. As Magritte interjected us with his painting, we can wonder why we were tricked into seeing a pipe when the canvas only contained pigments. Or more to the point, why do we see a cell, a vesicle, etc. from the list of numbers that comprise a scientific image? Answering this question actually triggers an ad hoc analysis of the image that we can use later. For instance, look at Figure 3.1c. This is a transmitted light image of two floating cells in contact, one of them is held by a micropipette. We see with a decent precision the contour of the two cells, and the area of contact between the two. Why is that so? Why do we see cells? They are imaged by transmitted light and feature a roundish shape. When the focus is set at their equator, it refracts the light away from the objective. Therefore, the equator appears as a dark outline around the cell that dominates out-of-focus artefacts. Our eyes perceive objects thanks to their contour very well. Since the cell contour is dark and well contrasted on this image, it drives us to individualise and localise two cells. A segmentation approach could emulate this and fit a closed contour around each cell by minimising the sum of the pixel values it traverses. Transmitted-light images of cells are typically very rich in texture, which can compromise this approach. To increase the robustness of the algorithm at the cost of accuracy, we can impose the cell contour to be a circle with its radius and centre being free parameters (Figure 3.1d).

Let's focus on Figure 3.1a again. The blebbing cell is imaged again with transmitted light at high resolution. The majority of the image content is made of greys and there is no clear contour line that tells where the cell is. However, we can tell very easily what part of the image is the cell and the background. How so? Though we capture the shape of the cell intuitively, analysing why is not immediately obvious. Grey levels are likely not the features that dragged our eyes to the cell here, as the pixels inside the cells sample many very different values. We actually see a cell here because of the texture inside it. The cell content refracts light in all random manners and leads some parts of it to focus more or less light to the objective than others. This generates pixel values that vary rapidly in space, compared to, for instance, the background which varies very little. A simple segmentation technique emulating this approach can consist of looking at the standard deviation of the values in a neighbourhood of the right size around each pixel. The cell segmentation is then built by taking all the pixels with a large standard deviation.

Fluorescence images are often easier to grasp. Classically, fluorophores are attached to specific proteins or to chemicals that target specifics components of the cell. Fluorescence images are therefore desirably sparse and specific – specific because they record signal coming only from the targeted component, and sparse because these targeted components are hopefully not spread over the whole field of view. Bright objects immediately drag our eyes over a black background. Segmentation strategies are simple in principle: pixels with large values are part of the object, whereas low values point to the background. A simple intensity threshold will classify pixels as one or the other.

Though trying to emulate our vision might sound appealing in all situations, many powerful segmentation techniques are not that intuitive. Some of them are not intuitive at all and root in the darkest and farthest corners of mathematics and computer science. Plus the simple ideas we surveyed are often found feeble against the noise that cripples most of the images we deal with. The eager analyst will stay informed and cultured on analysis techniques that can be found in successful publications as well.

3.1.3 A Strategic Approach

The exercises above might lead us to think that all successful quantifications are made by carefully analysing the images that are offered, and selecting the best algorithm that can extract objects or information from it. But you can be much more efficient if you start deriving an analysis process early, ideally before image acquisition. Most likely, you hold this book in your hands because your scientific projects involve image analysis. But they have a larger scope, and you will probably also be the one that performs the image acquisition, working on the microscope or on some imaging device. This means that you have full control over the images that you will analyse and can optimise them for a specific task. The right image can greatly simplify a complex analysis problem so it is best to think about the image you need sitting in front of the microscope rather than on the computer. Think about it for a minute. You don't need subcellular resolution to count the number of confluent adherent cells. The problem can be made trivial with a 5x objective and DAPI staining. It can be non-trivial in transmitted light microscopy. This is not the scope of this chapter, but think about strategy for analysis.

3.2 EXTRACTING OBJECTS

3.2.1 Detecting and Counting Objects

There are many applications in Life Sciences that require only a minimal amount of information on objects prior to moving to more elaborate quantification. Getting the count of the relevant objects and their respective x, y, z position is enough to yield all possible statistics on their relative spatial arrangement and their abundance. These numbers do not require generating the precise object contours. We therefore do not have to turn to potentially complex segmentation techniques. The only output we need from the image is the number of objects it contains and a list of x, y – and potentially z coordinates, one tuple for each object. So counting objects is in theory one of the most elementary forms of quantification.

Automatic counting of objects is almost always done on images where objects appear bright over a dark background, ideally flat. Fluorescence microscopy provides good examples of such images. In the following, we require that the objects we are interested in are all compact, roundish clumps of bright pixels. Such objects are typically nicknamed 'blobs' or 'spots'. For instance, nuclei stained with DAPI or Hoechst dyes make such objects. The same goes for cells stained with a cytoplasmic fluorescent dye observed at low magnification. At low resolution, cells stained for their cytoskeleton are also eligible, as the low resolution will blur the filamentous structures together in a single, though textured, object. Also, vesicles or any punctate subresolved structures will do.

3.2.1.1 Local Maxima and Image Noise

The first a priori information we can use to derive a detection strategy is that objects are made of bright clumps of pixels. One clump makes one single object; two objects are separated by the background and their pixels do not touch each other. In ideal cases, objects would have a pixel intensity that looks like a smooth peak. We can detect such objects by simply searching for local maxima. The pixel at the peak has an intensity that is larger than all the pixels in its immediate neighbourhood. So the number of local maxima gives us the number of peaks, therefore objects, and their location gives us the object positions.

In practice, the images we deal with in life sciences are almost always corrupted by noise [2]. In our context, noise is a generic term that encompasses the effect of all the deterioration sources impairing the acquisition of images. It is caused by non-ideal imaging devices (including the microscope stands, cameras, etc.) and limited photon collection (caused by weakly fluorescent samples, brief exposure times, low excitation intensities, etc.). Our subjective notion of good image quality correlates with a low level of noise. On images, noise shows up as undesired fluctuations around the mean. Ideally, a pixel that measures over time a fixed portion of a sample emitting a constant light intensity should display a constant intensity value as well. Because of noise however, this intensity will fluctuate randomly with a certain amplitude. For still images, this shows up as random spatial variations. The image of a sample emitting uniform light intensity (a flat object) will not display a single uniform pixel value, but will show random variation from one pixel to the next. This is very detrimental for our initial detection strategy, as noise might induce, for some pixels, an intensity that might be larger than all its neighbours. It will be counted as a local maximum, thereby generating a spurious detection. Plus, objects are very seldom made of smooth peaks. There might well be secondary maxima inside a single object that will again create spurious detections. Also, some objects might not be well separated by the background, and a single maximum could be detected for several objects. To moderate these problems, particular tools use a variety of techniques.

ImageJ ships a tool to extract local maxima [3] of 2D images, which works just the way we described. For each pixel of the source image, it tests if one of its eight neighbours has larger pixel intensity. If not, it is a local maximum and a detection is created for this pixel. Noise and secondary maxima can be accommodated thanks to a *noise tolerance* parameter, which in effect helps suppressing secondary maxima. Contiguous regions are formed from pixels with intensities ranging from the global maximum minus the *noise tolerance* value. Then a single maximum is accepted per region.

3.2.1.2 *Image Filtering and Accurate Detection*

The main strategy to deal with noise prior to detection is image filtering. An adequate set of filters is used to generate a new noiseless image from the raw data. The choice of filter is of course critical to getting meaningful results. One of the simplest and most efficient filters is the Gaussian filter. It requires a single parameter, the standard deviation, σ, that controls the spatial extent of the filter. The filtered image is obtained by replacing each pixel value by a linear combination of the pixel values in its immediate neighbourhood, weighted by some function. In the specific case of the Gaussian filter, the weights are calculated from the Gaussian function of the distance to the central pixel. Therefore, the value of σ determines the size of the neighbourhood. Here is an example in 2D for $\sigma = 1$.

		b		
0.00	0.01	0.02	0.01	0.00
0.01	0.06	0.10	0.06	0.01
0.02	0.10	0.16	0.10	0.02
0.01	0.06	0.10	0.06	0.01
0.00	0.01	0.02	0.01	0.00

$$p'(i,j) = \sum_{a,b} w_{a,b} \times p(i+a, j+b)$$

You can note several things from the neighbourhood. The linear combination shows that you are averaging several pixel value together to compute a new one, which will efficiently mitigate the noise in the filtered image. The larger the value of σ, the bigger the neighbourhood, so we can infer that heavy noise can be dealt with by a larger value of σ. However, when calculating a filtered pixel value, you incorporate pixels that can be far away (with increasing σ). This will generate some blurring of a potentially sharp image, which is often an undesired effect. There is therefore an optimum to find for the value of σ.

As a side note, the filtering technique we just described, which relies on the linear combination of neighbour pixels, is much more general than just the Gaussian filter. You can find or design hundreds of filters, called linear filters, by just changing the values of the weights $w_{a,b}$. For example, the mean filter is obtained by settings all weights to 1/9th over a 3×3 neighbourhood (this is the *Smooth* command of ImageJ, but since it is advantageously replaced by the Gaussian filter, we won't linger on it). Other types of filters are nonlinear and are not made of a linear combination of pixel values. The median filter for instance is obtained by collecting, for each pixel, the list of neighbour pixel values, and retaining the median of the list.

The Gaussian filter is typically used with a small σ to efficiently remove the noise from any image. But we want to use it in a blob detection algorithm, and have it generate a desirable image where objects are represented by a smooth peak. Minus a proper demonstration, the value of σ must be chosen so that the extent of the filter matches the typical size of the objects you want to detect. If it is too small, the filter will be sensitive to secondary maxima or noise. If it is too large, peaks for several objects might be merged into a single one. So here you need to feed to the algorithm a piece of a priori information: the rough size of the objects you want to detect. Once you have the filtered image, you just have to search for local maxima as before, but this time on an image devoid of noise.

Though the Gaussian filter already provides an excellent choice of filter for our needs, it was shown that a better precision and robustness could be achieved with the Laplacian of Gaussian (LoG) filter [4]. Also known as the Mexican hat filter, it takes this last name from the shape of its profile. It is mathematically defined by

$$\Delta g = -\sigma^2 \times \left(\frac{\partial^2 g}{\partial x^2} + \frac{\partial^2 g}{\partial y^2} \right)$$

where g is the source image smoothed by a Gaussian filter of standard deviation σ. Its shape in 1D resembles a sharp peak surrounded by two smaller minima. Again, you need to specify a value for σ. Regardless of its mathematical details, its effect is to give a marked response for bright and roundish objects of a size $2\sigma \sqrt{d}$ (in diameter) where d is the dimensionality (2 for 2D images, 3 for 3D, etc.). Objects with a different size will also generate a peak in the filtered image, but with an intensity lower than for an object of a similar brightness and of the optimal size. It also has the advantage of being less sensitive against slowly varying backgrounds, and it has been proved that it is an optimal detector against noise for roundish objects [4].

You can take another step to refine the detection results. Indeed, there might be smaller and larger debris or other undesirable objects in your sample that can generate local maxima that you wish not to detect. Hopefully, they will be dim, or not of the right size. The local maxima in the LoG-filtered image have an intensity value larger for objects that are bright and of the adequate size ($2\sigma \sqrt{d}$), you can use as a quality of detection value. Each peak has a quality value, and you can reject peaks with a quality lower than a certain threshold. The specification of a threshold on peaks allows stating whether a peak is eligible as an object and control the accuracy of the detection. Because of the LoG filter sensitivity, a detection is considered genuine or spurious based on its intensity and rough size. The specification of a value for the filtered intensity threshold again requires some a priori information. Its actual value cannot often be intuitively related to the pixel intensity in the raw image.

On the other hand, it might be tricky to provide it by manual adjustment when trying to compare the number of detected objects between conditions, since it is subjective. It is best then to use a negative control image and a positive control image to come with a sensible estimate of the threshold.

Many software packages offer the tools needed to apply this detection technique. The FeatureJ package [5] proposes 2D and 3D versions of the LoG filter for Fiji/ImageJ. You can then use the *Find Maxima* command to find peaks on the filtered image. Still in Fiji/ImageJ, TrackMate [15] ships several spot detectors also based on LoG filtering, and the SpotTracker [4, 6] tool as well. The Spot segmentation module of Imaris [7] is based on this technique.

3.2.1.3 *Filters and Image Processing*

Let's step back a little and reflect on the image-filtering step we took. We used the Gaussian or LoG filter to generate a new image devoid of noise and with a desired smoothness. We then feed this new image to the local maxima detector, which yielded the location of the objects we wanted to detect. Filtering an image prior to detection is a very common and general step. There are numerous filters well established in the domain that can transform any raw image into something suitable for further analysis.

For instance, we already observed that dealing with transmitted, bright-field images of cells was not trivial. With this imaging modality, cells do not display a smooth peak over a dark background. We already reasoned that it would be better to inspect the value of the standard deviation or variance in a neighbourhood around each pixel. The standard deviation is expected to be low for pixels outside of cells (the intensity does not vary a lot from one pixel to its neighbour), and high for pixels inside the cells. Indeed, cell material quickly generates spatially varying intensities. The resulting image resembles a fluorescent image, and you are now back to a problem you can solve (Figure 3.2). The techniques that consist in transforming an image into another one, better suited for detection or segmentation, are grouped in a domain generally called image processing. Image processing differs slightly from image analysis in that its output is another image rather than discrete numbers.

3.2.1.4 *Automated Scale Selection and Wavelet Transform*

As we have seen, the LoG filter offers a robust detector that is sensitive to a certain size – or scale – in particular. You must know and provide this scale prior to detection. Suppose now that you have to deal with

(a)   (b)

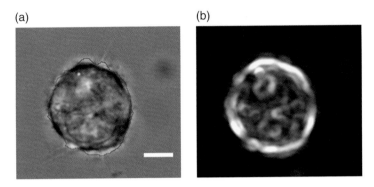

Figure 3.2 (a) L929 cell observed in bright-field overlaid with the segmentation results of the image in b. (b) Image from a. transformed with a 3×3 standard deviation filter. Scale bar: 5 µm.

spots of varying size. How could you detect them robustly without privileging a certain size over others? People have proposed using a combination of several LoG filters (or approximations), each tuned to a different scale, and applying them sequentially to the source image. To find spots and determine their size, you have to not only look for local maxima in x, y, z, but also find the filter that gave the largest peak value. The scale of this filter gives the most likely size for the spot inspected [8].

Another approach is based on an extension of the notion of filtering: wavelet transforms, whose technical details are by far out of the scope of this book. This mathematical framework extends the notion of filtering in that it allows us to directly interrogate the scale of features it highlights. A spot detector that accommodates varying scale and that relies on wavelets has been specially developed for biological images [9], and can be found in the Icy software package [10].

3.2.1.5 *Finding is Fitting*

So far we have focused on techniques that detect how many objects of interest are in the image, regardless of their exact shape, which is not analysed. We saw, however, that some detectors could be tuned to a certain object size, and that the detector could even obtain this size. Some techniques go a step further and can estimate the rough shape of detected objects, should it be a bit more complex than a blob, still without having to segment the exact object contour.

The object detector provided in the Kalaimoscope MotionTracker software [11] uses such a technique. It operates on 2D images and has an approach based on iterative peak fitting that can retrieve information beyond blob location and estimated radius. In summary, each bright object is fitted by one or a combination of several mathematical functions. The software uses the Lorentzian function, which resembles the Gaussian function but has some robustness properties interesting for fitting. Because the fit will operate on many pixels in a neighbourhood around the central pixel, the procedure is inherently robust against noise. A successful fit actually represents a kind of idealised object, devoid of noise and only represented by a sparse number of parameters: its peak intensity, position and extent. Interestingly, in Kalaimoscope the object detection step is conducted in an iterative way: The neighbourhood around the largest pixel intensity is fitted by a Lorentzian. If the fit is successful, it is stored for later analysis and the intensities returned by the fit are subtracted from the source image, effectively removing a peak from it. The brightest pixel is now located elsewhere and the fitting procedure is repeated, building a list of peaks unless the source image is left with only noise.

The peak radius, intensity and its accurate location are free parameters of the fit, so you get this information as a result. The current software implementation allows for peaks not to be perfect circles. It can accommodate peaks that have an ellipse-like shape with any orientation, which gives additional information. Plus, some objects might have a shape more complex than a single peak. They will therefore yield several successful fits that will be combined in a single object representation, depending on their proximity. The advantage of this technique is that it is a step towards a more accurate shape description of an object. The Lorentzian fits combination allows for obtaining an estimate of the object spatial extent, even if it is irregular, while the techniques above are limited to a radius. This also permits a more accurate computation of the total intensity inside an object, a useful measurement, which can be corrupted for irregular objects when estimated by just blob detection.

This method is made to work with objects that resemble an intensity peak or are made of a small combination of several of these peaks. Unsurprisingly this approach proved particularly successful for objects whose size is close to the microscope resolution limit [12]. The resolution limit will in effect blur together any sharp contours into smooth peaks, which can be well described by the fitting function. It is robust against noise, and allows for quantifications that require a rough object shape description. It does so without calling for more computer-intensive segmentation techniques, and was therefore successful in large-scale analysis [13].

Fitting approaches to detection are generic, and are also used as a building block in some algorithms. They are common when the image of an object to fit does not contain too many crisp features or when it resembles the fitting function. This can be the case when the imaged object is a size close to the resolving power of the chosen imaging modality, as already stated [13]. When raw images contain complex objects, and that complexity is not needed, image processing can be used to condense this information and generate an intermediate image with less content, and suitable for fitting. For instance, a modern cell detection technique [14] relies on clustering pixels into supervoxels prior to fitting the resulting condensed image by Gaussian mixtures.

3.2.1.6 Detection Techniques Recapitulation

We have surveyed several techniques for object detection. Their goal is to provide at least the number of objects in an image and their location. The techniques we presented all require the objects to be roundish and bright over a black background, should this kind of image be obtained through a special imaging modality such as fluorescence imaging, or be generated from other images by image processing techniques.

An efficient detector can be obtained by first filtering the image with the LoG filter then finding objects by looking for local maxima. The LoG filter allows for getting rid of noise, and provides a certain sensitivity to blobs of a specified radius. In practice, any software package that offers image filtering can be used to implement this detector, like MATLAB, R or any programming language supported by an adequate image-manipulation library. Non-programming software such as ImageJ often ships an LoG filter implementation [5] and a local maxima finder [3]. Some Fiji plugins are based on this technique, TrackMate for instance [15]. It also fosters the spot detector in Imaris [7]. More elaborate techniques are able to retrieve the approximate radius of detected objects, such as the spot detector of Icy [10]. Going one step further, Kalaimoscope MotionTracker [11] can deal successfully with objects of size close to the microscope resolution limit and return a rough shape description.

If you need to move beyond object number and location and require a precise description of the object shape, you have to move to another set of techniques, object segmentation. The aim of these techniques is to automatically retrieve an accurate representation of an object contour. They are often based on very different approaches and are best presented separately.

3.2.2 Automated Segmentation of Objects

As we said before, we use the term segmentation for techniques whose aim is to detect objects and get their boundaries, with an optimal accuracy. They are set apart from the detection techniques surveyed before in that the object contour is their focus. We rely on them when we want more than just object location, but also need their accurate shape. Also, they are made to deal with objects that differ greatly from the blobs defined in the previous section.

The image segmentation field is a very active one, and new algorithms and refinements are constantly produced. Even if we limit ourselves to bioimages, the sheer number of available techniques is overwhelming. This likely reflects the fact that there is no good-for-all method, and that a particular question often requires the development of a customised segmentation technique that exploits specifically the a priori knowledge of the application. Nonetheless, it is possible to root this multiplicity of techniques into several basic concepts that we will now survey.

3.2.2.1 Thresholding, Pixel Classification and Object Labelling

As we noticed above, fluorescence images are particularly convenient for analysis. A proper staining strategy will bind fluorophores to the objects of interest. In fluorescence imaging, the pixel value linearly encodes the density of fluorophores at the pixel location on the sample. So the object presence is directly reflected in the individual pixel values: bright pixels (high pixel values) are likely to belong to a stained object, dark pixels (low pixel values), not (Figure 3.3a).

The simplest way to exploit this is to set a threshold on pixel intensity that will partition the image pixels in two classes (Figure 3.3b). If a pixel has a value below the threshold, it is said to belong to the background. If its value is above the threshold, it belongs to an object, or foreground. We can generate a new image from this threshold, for which pixels will receive a value of 0 if the source image has a pixel value below the threshold, and 1 otherwise. Since the new image is made only of two values, it is called a binary image or a mask (Figure 3.3c). They are typically represented in black (for pixel values equal to 0) and white (for pixel values equal to 1).

The mask considerably condenses the raw information from the source image: all the intensity information is discarded in favour of black or white values. Still, this is not enough to extract objects. To do so, adjacent or connected pixels are searched and a unique identity is assigned to all the foreground pixels that are 'touching' each other.

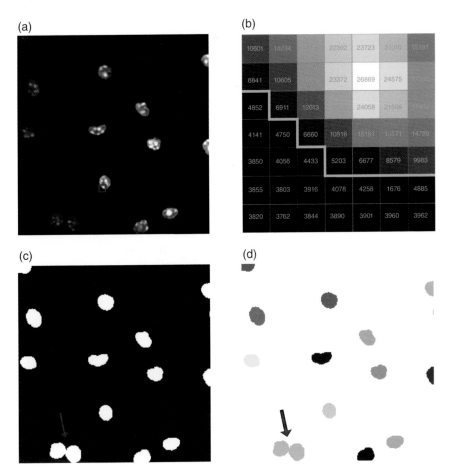

Figure 3.3 (a) Macrophages fixed and stained with DAPI observed in fluorescence. (b) Close up version of the red box in a. The green line delineates a contour obtained by putting a threshold on intensity at 5000. (c) Binary mask obtained by thresholding intensity at 5000. Note that the masks of the two cells indicated by the red arrow are touching. (d) Objects after labeling. Different objects are indicated by different colors. The two touching cells generated a single object. Sample courtesy of Thibault Rosazza and Eric Prina, Institut Pasteur, Paris, France.

This step is called object labelling (Figure 3.3d). The set of connected pixels with a shared identity forms an object, from which a contour can be extracted and quantification performed. Of course, the final objects depend on how you define 'touching' or connected pixels. In software packages, you will sometimes find the terms '4-connected' or '8-connected'. The 4-connectivity defines two pixels as connected if they are in a 2D plane directly east, south, west or north of each other, excluding the diagonals.

The 8-connectivity includes the diagonals, still in a 2D plane. The 6-connectivity and 26-connectivity extend these definitions in 3D. In practice, the shape of bulky objects does not depend too much using 4-connectivity versus 8-connectivity or their 3D equivalent.

The *Analyse particles* command of ImageJ [16] automates all these steps. It takes a thresholded image or a mask image and outputs objects as several regions of interest. It also offers basic filtering of objects based on their size and roundness, but is limited to 2D objects. The CellProfiler software package [17] has an *Identify primary objects* that is also based on thresholding, but offers far more refinements, some of which are considered below. CellCognition [18] specialises in HCA/HCS and includes some tools based on thresholding. Commercial software that focuses on fluorescence images also extended this principle, for instance Definiens [19], Volocity [20] or Imaris [7].

3.2.2.2 *Automated Threshold Determination*

This approach is great in that we are using an a priori knowledge to properly condense information thanks to a simple thresholding on pixel value. However, it is flawed by subjectivity, since we set the threshold manually. For instance, objects with smooth borders will yield a mask that depends greatly on the precise value of the threshold. A small variation around a value that efficiently segments objects might greatly affect their size. Indeed, pixels on the border of the objects will have intensities that vary roughly continuously. They may switch from one class to another with even small variations on the threshold value. This is an inherent problem with thresholding: it is a rather brutal way of condensing information (pixels become black or white) and we lose detail.

This may impact negatively the rigor of the analysis if we set the threshold ourselves separately for each image we analyse. If thresholding is going to generate some systematic error on the estimate of the shape of an object, it is better if this error is the same for all images of a dataset. The solution is to turn to algorithms that automatically determine a threshold value for an image, based on the image content. There are a large number of them: the *Auto threshold* plugin of ImageJ [21] lists 16 of them. They all exploit a priori knowledge on the image histogram. For instance, the triangle algorithm [22] operates directly on the shape of the histogram. It looks for a peak on one side of the histogram and searches for a second one on the other side, then finds what pixel value separates them. This will fail if there are not two peaks, or if they are not arranged as expected. Otsu algorithm [23] assumes that the histogram comes from two classes of intensities, and finds the threshold by minimising their combined variance. They all aim at being relatively robust against image content variability. A practical approach would be to pick a thresholding algorithm that works for a subset of images, and to stick to it for the whole dataset.

3.2.2.3 Morphological Operations

As for object detection, noise can considerably alter segmentation results. If the fluctuations in intensity are large, a pixel belonging to the background may have a very large pixel value. If this value is above the threshold, it will be classified as an object. Conversely, a pixel belonging to the foreground might be classified as a background pixel for the same reason. Noise then degrades our ideal mask image in that it adds isolated pixels that are not classified correctly.

Of course, we can deal with noise as presented above, by pre-processing the image so as to denoise it. Another approach is to rely on morphological operations [24] to clean a noisy mask. Under this term, you can find numerous image processing algorithms, well suited to post-processing a binary mask, though they can be extended well beyond binary images. Their description and capabilities extend far beyond the scope of this chapter, and we will quickly look at just four of them (Figure 3.4). The *Dilation* operation acts as if you were repainting foreground objects with a brush, called the *structuring element*. Every foreground pixel neighbourhood is replaced by the structuring element shape. In effect, this looks as if the foreground object would dilate by getting a sort of coating on their boundaries, set by the structuring element shape (Figure 3.4b). *Erosion* is just the converse operation. Here everything works as if you had an eraser of a certain shape. Foreground objects are shrunk up to an extent set by the structuring element (Figure 3.4c). The *Close* (Figure 3.4e) morphological operation consists of the dilation and erosion operations applied in succession. The *Open* operation (Figure 3.4f) also does this, but in the reverse order. In practice, these last two operations can be used to efficiently clean a binary mask mildly corrupted by noise and efficiently remove small spurious objects and smooth the contour of relevant objects, still respecting the rough size of objects in the initial mask.

ImageJ (and therefore Fiji) includes a collection of tools to operate on binary masks, including the four operations cited above. They can be found in the *Process > Binary* menu. *MorphoLibJ* [25] is an ImageJ library that greatly extends these features, notably adding the capability

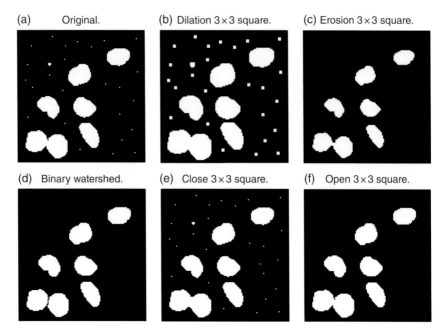

(a)   Original.   (b) Dilation 3×3 square.   (c) Erosion 3×3 square.

(d) Binary watershed.   (e) Close 3×3 square.   (f) Open 3×3 square.

Figure 3.4   Some morphological operations. (a) Synthetic example mask generated by taking portions of figure 4a. (b), (c), (d), (e), (f). Results of respectively the dilation, erosion, watershed, close and open operations on the source image in a. Dilation, erosion, close and open operations used a 3×3 square structuring element.

to deal with 3D images. Icy has a plugin called *MorphoMaths* that regroups the morphology tools. Programming software packages often ship a morphology framework. MATLAB morphology capabilities are included in the Image Processing toolbox.

3.2.2.4 The Crux of Touching, Distinct Objects

A classical problem with object labelling is touching objects. Since thresholding strategies involve a binary mask, as soon as two distinct objects touch in the mask, they will be labelled as a single large object. This will negatively affect the accuracy of almost anything you can quantify from the segmentation. You will not get the right number of objects; some of them will appear too large, etc. This is what has happened in Figure 3.3c. Two nuclei each belonging to a distinct cell have their mask too close, and some of their pixels are neighbours. The labelling algorithm therefore iterates them at once, and outputs a single object made of two blobs (Figure 3.3d).

Our eyes immediately identify the faulty object, and we can easily draw a line where the big object should be cut to yield two correct object masks. We can do so because we know that the expected shape of the correct objects is roughly round, and we clearly see a pinch in the middle in the big object. Mathematical morphology provides a technique that emulates this, and can make a cut where the big object is pinched, using what is called the watershed technique (Figure 3.4d). Under this term is hidden a shortcut through several processing steps whose start and end points are a binary image, and which we will not detail. It belongs to the morphology category, operates on binary images, and you will often find it at the same place in software packages. For instance, in ImageJ it lives in the *Process > Binary* menu. In practice, the watershed technique will deal successfully with touching objects when they are roundish and when they are touching through a slim contact. This means that it will work poorly for cases that deviate from this. In particular, if the objects you are trying to separate are very elongated, the watershed will over segment them, leading this time to many small spurious objects.

3.2.2.5 Segmentation of Objects from Their Contours

The techniques we have looked so far deal with so-called *object images*, where objects appear bright as a whole over a darker background – typically cells, nuclei and organelles stained in their volume. They would not work well for *edge images* where the object *contours* are bright. You can get this kind of image, for instance, if you stain the cell membranes (Figure 3.5a) or in phase contrast or as an output of a pre-processing step. If you apply directly the thresholding techniques described above, it will yield the contours as single objects, rather than the actual objects they delineate.

A technique called morphological segmentation exploits the watershed technique [24] that we used on binary images, but this time on greyscale images. The underlying idea is to deal with the image as an elevation map (Figure 3.6). Large pixel values represent high altitude and small pixel values low altitude. In this view, our image resembles several valleys, corresponding to the objects of interest, separated by steep ridges (Figure 3.6a). Let's now imagine that water would percolate from the ground (Figure 3.6b). Its level would increase, inundating each valley. Water reaches the valleys with a high altitude later than others, but as long as the water level does not reach the ridge, waters from separate valleys do not mix. When a basin meets another one on these ridges, a dam is traced to separate objects. Hence the name, watershed.

(a)　　　　　　　　　　　　　　(b)

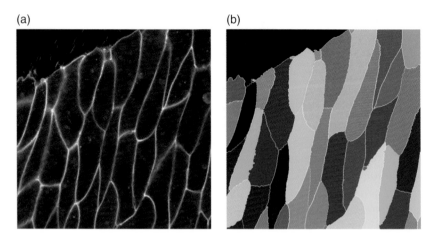

Figure 3.5 (a) Epithelium of the neural plate of *Xenopus laevis* visualized by utrophin-GFP. (b) Results of segmentation using the Morphological Segmentation Fiji plugin. Image courtesy of Jakub Sedzinski, UT Austin, USA.

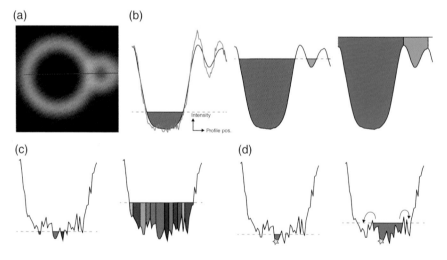

Figure 3.6 The watershed segmentation technique. (a) Synthetic noisy image depicting two compartments delineated by bright and noisy contours. (b) Watershed technique illustrated on an intensity profile through the red line in a. The profile was filtered to appear smooth. From left to right the "water level" rises, flooding compartments. The object contours are set where waters from two basins meet. (c) Actual watershed on a non-filtered profile. Intensity fluctuations due to noise generate several small basins that lead to numerous spurious objects. (d) Marker-controlled watershed. Water flows only from a discrete numbers of sources set by markers. As the water level rises, it floods the small, noisy basins. In the end, this technique generates one object per marker.

Objects are made of all the pixels that are flooded from the same basin. Notice that this technique gives contiguous objects that fill the image, and that the background is considered as any other object.

Several variations of the algorithm have been developed, and made to offer different strategies to deal with oversegmentation. Indeed, noise can cause spurious ridges to appear at low intensities, creating unwanted divisions of the real object. On the noisy image of Figure 3.6a, the direct application of the watershed algorithm gives rise to several hundred objects instead of the expected three (Figure 3.6c). Noise generates small spurious valleys. All these numerous local minima will give rise to small objects as the water level rise, instead of a single large one. Pre-processing is an option, but other strategies offer to control flooding by imposing seeds for objects. If we carry on with our water analogy, seeds would be sources, the only places from which water would flow. A valley devoid of a source would be flooded by the water coming from the adjacent seeded valley, when it passes the lowest ridge (Figure 3.6d). This approach leads to having one object per seed, so properly defining the seeds is essential. But it deals efficiently with noise, because spurious valleys are flooded by the main seeded one. Some implementations let you define the seeds yourself [26], for instance the *Marker-controlled Watershed* plugin of the *MorphoLibJ* library [25]. Others automate several tasks, such as the *Morphological Segmentation* plugin. It finds the seeds automatically by looking for the regional minima in the image with a specified tolerance to get rid of spurious edges (Figure 3.5b).

3.2.2.6 Region Growing

The seed-based watershed technique we have just skimmed over is an iterative process that could be very crudely summarised like this:

1. Start from object seeds at local minima, regional minima or imposed positions, and put the water level at the seeds' level.
2. Raise the water level by a small increment and discover all pixels with a value lower than the water level.
3. All the discovered pixels that are not yet labelled are incorporated in the adjacent object.
4. Repeat from step 2 until the water level is larger than the maximal pixel value.

In effect we are growing objects from seeds. They grow as large as they can, until they meet on the top of a ridge.

The watershed technique is well suited for edge images but we can extend its concept to object images. Here we attempt at having a method that can deal with objects stained roughly uniformly in their volume. We need to alter the way we stop growing objects, since on object images there are no ridges to stop their growth. The outline of such an algorithm would be like this:

1. Start from object seeds.
2. Discover all pixels directly adjacent to a growing object.
3. All the discovered pixels that are *similar* to the pixels in the adjacent object get incorporated in it.
4. Repeat from step 2 until there are no more pixels to discover.

Of course, we need to specify what we mean by *similar*. For instance, you could decide that a discovered pixel value p must be close to the mean pixel values o_i of the object to be incorporated [27]. With a specified tolerance Δ:

$$p \text{ is incorporated if } | \text{ mean}(o_i) - p | < \Delta$$

At each iteration, the mean is updated with the new pixels that were incorporated in the previous step. The *Fast marching* algorithm can be seen as an extra layer of refinement over region growing, and that gives specifics on step 2. The algorithm prioritises what pixels are discovered and processed first, based on the image data. Again, the growing process resembles a region being flooded: the water flows first in areas of flat intensities. The *Level sets* plugin of Fiji [28] ships an implementation of the Fast marching algorithm, along with another one we will detail later. Its starting point is a greyscale image annotated with point seeds. The Fast marching algorithm will flood objects from these seeds, and stops at object borders where a specified difference in intensity is met.

So far, we have assumed that images were made of bright objects over a darker uniform background. Region growing techniques are good at handling images containing objects with very different mean intensity, or where the background varies a lot. In such cases, an object in a dark part of the image might have pixel values below the background in a bright part of the image. Applying a global threshold as we did above will utterly fail. When the background varies a lot, it is impossible to pick a threshold value that will work everywhere. Several image-processing techniques exist to correct for a non-uniform background [29], or you can choose to rely on local thresholding [30]. Region growing techniques are a viable alternative, as they only consider the local intensities around an object. They are less affected by a non-uniform background, provided it does not alter the stop criterion of the algorithm. They require seeds to be specified. This can be interesting for semi-automatic workflows; you might want to specify manually the objects you want to extract in the vicinity of many others. But this can be a major hurdle when trying to automatically analyse a large batch of images. In this case, a good strategy is to rely on an initial detection step to find the seeds automatically, based for instance on the detection algorithms surveyed in the previous section. A common approach is to detect seeds in a first channel (e.g. nuclei staining) and growing objects using a second channel (e.g. cytoplasm staining). This is particularly efficient when trying to segment confluent cells, and many software packages facilitate this process, for instance CellProfiler [17] or Definiens [19].

3.2.2.7 Deformable Contours

Region growing techniques have an important weak point. They depend on the object border to be perfect everywhere. If there is a tiny gap on the border, the flooding will leak out of through the gap, leading to an inappropriately large object. To prevent this, you could use the a priori knowledge on the object area (or volume), but simple region growing techniques cannot make use of such a constraint. Actually, they only use local intensity at the object border as constraints, which are used as a stop criterion for the region growing part. But sometimes you know more than this. You may be able to tell roughly how large is the object, or how smooth its contour is. Using this knowledge in a segmentation algorithm can result in far more accurate results, and avoid the leaking problem we just described. There is a full set of segmentation techniques that make use of these constraints, and that can be loosely grouped under the term deformable contours.

The body of work behind deformable contours is enormous and the literature is very rich. Very roughly, deformable contour algorithms can be grouped in two classes: 'snakes' or sometimes explicit parametric representation or simply active contours [31] and 'level sets' or implicit or geodesic active contours [32]. Practically for us, their input, output and process share similarities. As the name implies, these techniques take a contour as a starting point, and deform it iteratively until it accurately delineates an object's border. The input image can be an object image or an edge image, depending on the implementation. They use an iterative process: a contour is evolved from a specified starting shape, made to be pulled towards object edges, under several configurable constraints such as contour sampling, contour smoothness or curvature, preferred volume, etc. Their algorithmic details are fuelled by partial-derivative equations (PDEs) where image data takes the shape of a force that drives the contour towards edges, while curvature, volume, etc. constraints prevent this contour morphing into nonsensical shapes.

Implementations differ a lot. Snakes implementations are based on discretising the contour over a finite set of points, and minimising a sum of energy terms. There is an image energy term that drives the contour towards object edges, and terms that constrain the shape of the contour to be regular. In some ways, the technique that is used to segment as two circles, the two touching cells imaged in bright-field in Figure 3.1d, can be seen as an extreme version of snakes. That is, the image energy term is simply the sum of the pixel intensity sampled by the two circles, and the shape constraint translates by imposing the contour to be the union of two circles. Proper snake algorithms let the contour be an arbitrary polygon, more or less smooth depending on specified parameters. Level sets implementations on the other hand evolve a surface (a 3D surface for 2D images, a 4D hypersurface for 3D images), and the object contour is derived from the intersection of this surface with a particular plane. At least for us, this approach is hard to derive intuitively. As we noted in the introduction of this chapter, some techniques are the product of research in computer science and are imported into bioimage analysis. We will simply keep in mind that snakes are typically faster but often limited to simple topologies (a single, convex object, but not necessarily closed) and the output depends sometimes strongly on the specified initial contour, while level sets can handle complex topologies (several objects, possibly with holes) at some extra computational cost. Recent work however is made to temper the drawbacks of each approach.

Several software packages use deformable contours in their processing pipelines as a means of refining the segmentation obtained in other ways. This is the case for instance for the bacteria specialised tools MicrobeTracker [33] and its later inception Oufti [34]. In practice, these techniques excel at dealing with noisy or split object borders. For instance, Figure 3.7a shows a neural tube stained for cell membrane. But the intracellular volume is noisy with a mean level close to the membrane value. Also, some of the membranes have gaps caused by imperfect staining. A region-growing technique like the morphological segmentation we have used above completely fails (Figure 3.7b). A low tolerance to noise produces oversegmentation. A high tolerance misses most of the object edges, and the segmentation is made of a small number of objects made of many cells. A compromise between these two extremes gives rise to the two problems. Since there are gaps in some membranes, it is impossible to properly segment cells with such gaps. Snakes on the other hand can bridge over these gaps. They need to be initiated with a starting contour (in Figure 3.7c, an ellipse), and will evolve it until the object is properly segmented (Figure 3.7d). Fiji ships a level set implementation

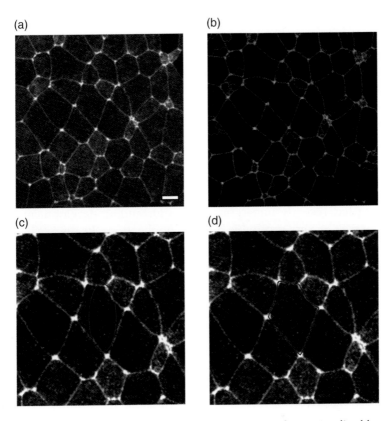

Figure 3.7 (a) Epithelium of the neural plate of *Xenopus laevis* visualized by utrophin-GFP. Scale bar: 10 μm. (b) Segmentation results using the Morphological Segmentation Fiji plugin. (c) Initialization contour over a target cell. Note that the top left membrane of the target cell displays a hole in its staining. (d) Results of segmentation with the Active Contour plugin of Icy. The technique could successfully bridge over the membrane gap. Image courtesy of Jakub Sedzinski, UT Austin, USA.

[28], that we used above because it included a region-growing-like initialiser. Snakes plugins can also be found, such as the E-Snake plugin [35]. Icy offers several plugins for deformable contours [36–39] implemented following recent advances in the domain [40–42].

3.2.2.8 Machine Learning and Object Segmentation

In the introduction of the previous chapter we marvelled at how quickly we could identify a pipe in a painting, without a priori context. Our enthusiasm at how fast we can analyse this image has to be relativised by the fact that we trained for a very long time before being able to do so.

You might even have seen this painting before and already knew its meaning. Although our visual perception is much more complex, we could make an analogy with a classification process. We classified the object in Magritte's painting as a pipe because we have been previously exposed to many pipes and told that they were pipes. So we had in our memory a class of objects that are pipes and the example objects in this class covered most of existing pipes. Facing the Magritte painting, we are presenting a new object that displays features that match the pipe class better than any other class we know of. So we identify it as a pipe. This very loose sketch of a human capability has an analog in computer science, in a domain called machine learning. Research in machine learning has recently undergone some outstanding developments, fostered in part by very profitable business applications. Its outreach affects many of the aspect of our everyday life, from product recommendations, predictive diagnosis to person identification, speech recognition, etc. Apart from being a research field by itself, it is used by scientists as a tool for many applications, including object segmentation.

Machine learning is about training a computer by providing it with data items, all categorised in a number of classes. From this classification, a machine learning algorithm extracts features from the data and builds a classifier that will allow it to make predictions. In a second step, new unclassified data is presented to the algorithm, and it assigns it to the right classes, using the classifier it built in the training phase. In machine learning object segmentation, the training phase consists of the user manually annotating patches of one or several images with the right class. After that, the algorithm is expected to retrieve the right segmentation on all images. An example of such a process with minimal configuration is presented in Figure 3.8a. It pictures cells imaged with a transmission electron microscope. One can distinguish roughly three classes of objects: mitochondria, plasma membrane and cytoplasm. These classes have marked differences in their texture, rather than just their intensity. Mitochondria are darker on average than the rest of the image, but using a plain thresholding approach leads to very unsatisfactory results, even after intensive binary mask post-processing. What distinguishes these three classes is not solely intensity. Though mitochondria are darker, they also display layers of alternating intensities. Similar layers can be found in the plasma membrane, but they are more spaced. The cytoplasm has roughly the same mean intensity as in the membrane, but it is constellated by ribosomes and irregular Golgi. These differences in the texture of patches can be picked up numerically by certain filters. We discussed, in the previous section on detection, the

(a)

(b)

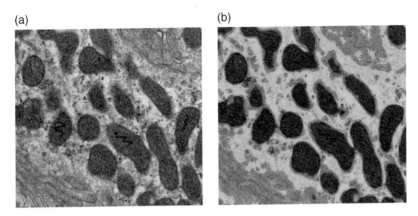

Figure 3.8 (a) Transmitted electron micrographs of the basolateral part of enterocyte cells in the guinea pig colon. The color overlay shows manual annotation of three parts of the cells: mitochondria (red), cytoplasm (green) and basolateral plasma membrane (purple). (b) Results of the supervised machine learning segmentation, using the Trainable Weka Segmentation plugin and the input classification depicted in panel a. Image courtesy of M. Sachse and E.T. Arena, Institut Pasteur, Paris, France.

importance of filtering and image processing. There exist many filters that are able to put numbers onto our crude description two sentences above. For instance, the LoG filter can tell whether there are some punctate structures in a patch. The Gabor filter can tell whether there are some elongated structures, etc. A combination of filters provides the numerical features that the machine-learning algorithm will feed on for the classification. Of course, the choice of the feature set has a major impact on the segmentation result. If all the features you select are common for two classes, they will be conflated. For instance, notice that the segmentation results on Figure 3.8b obtained with a default set of features is not perfect. Most mitochondria are coated by the membrane class, though we trained the latter only on plasma membrane. This is probably because the mitochondria membrane image and the plasma membrane image share texture similarities. Similarly, some of the Golgi found in the cytoplasm as labelled as membrane on the segmentation.

Machine learning based segmentation tools are particularly useful when the texture of objects are apparent and well resolved. This is why it is very common to see these tools used for electron microscopy, transmitted light images or histopathology images, and maybe a bit less often in fluorescence images. They are extremely useful for batch processing. Provided that all images of a dataset are consistent, the classifier built in the training phase, possibly on a single image, can be applied to the whole dataset at once. This can be computer intensive, but the effort required of the user remains modest. There is a whole world of algorithms behind the term 'machine learning' that we have not broached. It can be hard to understand why they work or, more importantly, what to do when they do not yield a correct segmentation. Some of these algorithms operate as a black box and just provide an answer without any clue of how they reached this answer. A sensible approach is to stick to picking the features that best tell apart the classes you see. The careful inspection of the image and the ability to decide what it is in the image that makes an object class different from all others are useful skills, as we noted in the introduction. It has to be supplemented with some knowledge of what are the features or image filters that will highlight these differences.

The open-source software implementations of machine learning object segmentation often rely on existing toolboxes developed within the academic domain of machine learning. Fiji ships a plugin for machine learning and object segmentation, based on the Weka machine learning toolbox [43]. The Fiji SIOX plugin [44] only operates on true colour images (e.g. as in histopathology images), and though it is not truly based on machine learning algorithms, it runs similarly. Ilastik [45] is software built mainly on machine learning, and it expands its range of applications compared to what is described here. In Icy, the Rapid Learning [46] plugin relies on the RapidMiner library. The Texture Segmentation plugin [47] implemented a specific set of features combining colour and texture to address in particular histological images [48]. Several commercial softwares implement the process we described here. The inForm [49] commercial software that specialises in anatomical-pathology applications also relies on machine learning. Definiens Tissue Studio [19] is another example of commercial software in a similar domain. Both also work with fluorescence images. As we said, machine learning is a discipline with many applications. Here we have just described one, where it is used for object segmentation after a training phase. Bioimaging has several other applications for machine learning. For instance, CellCognition [18] specialises in HCA/HCS applications and uses machine learning to classify objects that have been segmented otherwise.

3.3 WRAPPING UP

The goal of this chapter was to survey basic segmentation techniques, and to point to tools that make them accessible in practical situations. There are many other software tools, open-source and commercial, other than those summarised in the Table 3.1. As said in the introduction, it is more important to have a strategy for the questions you want to address.

Table 3.1 Commonly used Bioimaging software for quantification.

Image Analysis Package	Open Source - Manufacturer	Applications	Reference link
Acapella	No - Perkin Elmer	High Content Analysis, often used for phenotypic screening.	http://www.perkinelmer.co.uk/category/image-analysis-software
Amira + Avizo3D	No - FEI	Visualization and quantification of 2D and 3D data, Electron Microscopy data in particular.	https://www.fei.com/software/avizo3d/
BioImageXD	Open source - Universities of Jyväskylä and Turku in Finland, MPI- CBG, Germany	3D quantification and segmentation software based on the ITK and VTK libraries.	http://www.bioimagexd.net/
BisQue	Open source - UCSB, USA	"Store, visualize, organize and *analyze images* in the cloud". Collaborative, database oriented application with several analysis modules.	http://bioimage.ucsb.edu/bisque
Big Data Viewer	Open source – MPI-CBG, Dresden Germany	Specialized application for the interactive investigation of large, multi-view images. Used in particular with light-sheet microscopy images.	http://imagej.net/BigDataViewer
CellProfiler & CellProfiler Analyst	Open source – Broad Institute USA	"Software designed to enable biologists without training in computer vision or programming to quantitatively measure phenotypes from thousands of images automatically." CellProfiler analyst extends CellProfiler with the tools required to explore image-derived data.	http://cellprofiler.org/

(*Continued*)

Table 3.1 (Continued)

Image Analysis Package	Open Source - Manufacturer	Applications	Reference link
CellCognition	Open source - IMBA Vienna	Framework and standalone application for fast and cross-platform image analysis of fluorescence time-lapse microscopy in the field of bioimage informatics.	http://www.cellcognition.org/
Definiens	No - Definiens	Platform for automated image analysis, using an original development framework. Focuses on tissue analysis.	http://www.definiens.com/
Fiji/ImageJ/ ImageJ 2	Open source – Worldwide but led by Wayne Rasband (Ex NIH, USA) and Curtis Rueden, UW-Madison, LOCI, USA	ImageJ is an open source Java image-processing program. ImageJ has a strong, established user base, with thousands of plugins and macros for performing a wide variety of tasks. Fiji ships ImageJ and fletches it with many plugins and an update mechanism for them. ImageJ2 is the successor of ImageJ.	http://imagej.net/Welcomehttp://fiji.sc/ http://imagej.net/ImageJ2
Icy	Open source - Institute Pasteur Paris	"Open community platform for bioimage informatics." Application and website for image processing and analysis in bioimaging. Has a plugin framework with updates and ratings, graphical programming for the quick development of custom solutions and many more features.	http://icy.bioimageanalysis.org/

Name	Source	Description	URL
Imaris	No - Bitplane, Oxford Instruments	3D and 4D real-time interactive data visualization and analysis. Has a modular design, with modules for specific applications in Cell Biology, Neuroscience and Developmental Biology.	http://www.bitplane.com/imaris/
ILastik	Open source - EMBL Heildelburg Germany and HHMI Janelia Farm USA	"*ilastik* is a simple, user-friendly tool for interactive image classification, segmentation and analysis."	http://ilastik.org/
Image-Pro Premier	No - MediaCybernetics	2D and 3D Image analysis software that can be extended with image acquisition capabilities. Is not limited to Life Sciences	http://www.mediacy.com/imagepropremier
ImageTool	Open source - The University of Texas, USA	A free C++ image processing and analysis program for Pathology and fluorescence 2D images for Windows platforms	http://compdent.uthscsa.edu/dig/itdesc.html
Kalaimoscope	Open source - MPI-CBG Dresden, Germany	Sub resolution object tracking in 2D. Strong focus on grid computing to accelerate the analysis large screen datasets.	http://www.kalaimoscope.com/product.html
Leica LAS	No - Leica Microsystems	Processing and quantification of 2D and 3D images acquired on Leica microscopes using Leica acquisition software.	http://www.leica-microsystems.com/home/
MetaMorph/ MetaExpress	No - Molecular Devices	Modular software often used to control microscopes, and fully fledged with analysis modules.	https://www.moleculardevices.com/systems/metamorph-research-imaging/metamorph-microscopy-automation-and-image-analysis-software

(Continued)

Table 3.1 (Continued)

Image Analysis Package	Open Source - Manufacturer	Applications	Reference link
NIS Elements	No - Nikon Instruments	Processing and quantification of 2D, 3D and super-resolution images acquired on Nikon microscopes using Nikon acquisition software.	https://www.nikoninstruments.com/en_GB/Products
BIG	Open source - BioImaging Group, EPFL Switzerland	New algorithms and mathematical tools developed by the BIG team for advanced processing of biomedical images *e.g.* SpotTracker, Deconvolution Lab.	http://bigwww.epfl.ch/
Trackmate	Open source - Institute Pasteur, France	One of the single-particle tracking tools distributed with Fiji. Has a focus on usability and pluggability.	http://imagej.net/TrackMate
Vaa3D	Open source - Allen Institute of Brain Science, USA	"A Swiss Army knife for exploring big big image data." Extracts complex surface objects from images, and performs comprehensive analyses such as brain connectome mapping.	http://www.alleninstitute.org/what-we-do/brain-science/research/open-science-research-tools/vaa3d/
Visiopharm	No - VisioPharm.	A quantitative digital pathology solution providing flexible, powerful quantification of tissue properties in a fast, objective and reproducible way.	http://www.visiopharm.com/
Zen	No - Zeiss Microscopy	Processing and quantification of 2D and 3D images acquired on Zeiss microscopes using Zeiss acquisition software.	http://www.zeiss.com/microscopy/en_de/home.html

Focus first on what quantifiable information is required to reach a scientific conclusion, pick an analysing technique that can yield it, find or write a software tool that implements it, and finally acquire images that can be analysed with such a tool. A quantification strategy has to be designed from end to start.

3.4 REFERENCES

1. Tinevez, J.-Y., et al., *Role of cortical tension in bleb growth*. Proceedings of the National Academy of Sciences, 2009. **106**(44): p. 18581–18586.
2. Waters, J.C., *Accuracy and precision in quantitative fluorescence microscopy*. J Cell Biol, 2009. **185**(7): p. 1135–48.
3. Rasband, W., *Find Maxima… in ImageJ*. p. In ImageJ menu, Process > Find Maxima… http://rsb.info.nih.gov/ij/docs/menus/process.html#find-maxima.
4. Sage, D., et al., *Automatic tracking of individual fluorescence particles: application to the study of chromosome dynamics*. IEEE Trans Image Process, 2005. **14**(9): p. 1372–83.
5. Meijering, E., *FeatureJ: A Java Package for Image Feature Extraction*, in *ImageJ and Fiji*. http://www.imagescience.org/meijering/software/featurej/.
6. Sage, D., *SpotTracker*, in *ImageJ and Fiji*. http://bigwww.epfl.ch/sage/soft/spottracker/.
7. BitPlane, *Imaris*. BitPlane AG. http://www.bitplane.com/imaris.
8. Lindeberg, T., *Feature Detection with Automatic Scale Selection*. International Journal of Computer Vision, 1998. **30**(2): p. 79–116.
9. Olivo-Marin, J.-C., *Extraction of spots in biological images using multiscale products*. Pattern Recognition, 2002. **35**(9): p. 1989–1996.
10. de Chaumont, F., *Icy Spot Detector*, in *Icy*. http://icy.bioimageanalysis.org/plugin/Spot_Detector.
11. Kalaidzidis, Y., M. Zerial and M.R. Alvers, *Kalaimoscope MotionTracker*. http://www.kalaimoscope.com/.
12. Rink, J., et al., *Rab conversion as a mechanism of progression from early to late endosomes*. Cell, 2005. **122**(5): p. 735–49.
13. Collinet, C., et al., *Systems survey of endocytosis by multiparametric image analysis*. Nature, 2010. **464**(7286): p. 243–9.
14. Amat, F., et al., *Fast, accurate reconstruction of cell lineages from large-scale fluorescence microscopy data*. Nat Methods, 2014. **11**(9): p. 951–8.
15. Tinevez, J.Y., et al., *TrackMate: An open and extensible platform for single-particle tracking*. Methods, 2017. **115**, p. 80–90.
16. Rasband, W., *Analyse Particles… in ImageJ*. http://rsb.info.nih.gov/ij/docs/menus/analyse.html#ap.
17. Kamentsky, L., et al., *Improved structure, function and compatibility for CellProfiler: modular high-throughput image analysis software*. Bioinformatics, 2011. **27**(8): p. 1179–80.
18. Held, M., et al., *CellCognition: time-resolved phenotype annotation in high-throughput live cell imaging*. Nat Methods, 2010. **7**(9): p. 747–54.

19. Definiens, *Tissue Studio*. http://www.definiens.com/research/tissue-studio.
20. Perkin-Elmer, *Volocity*. http://cellularimaging.perkinelmer.com/downloads/detail.php?id=14.
21. Landini, G., *Auto Threshold*, in *ImageJ*. http://imagej.net/Auto_Threshold. In ImageJ menu, Image > Adjust > Auto Threshold. Also ported to Icy by T. Provoost, in the Best Threshold plugin.
22. Zack, G.W., W.E. Rogers and S.A. Latt, *Automatic measurement of sister chromatid exchange frequency*. Journal of Histochemistry & Cytochemistry, 1977. **25**(7): p. 741–53.
23. Otsu, N., *A Threshold Selection Method from Gray-Level Histograms*. IEEE Transactions on Systems, Man, and Cybernetics, 1979. **9**(1): p. 62–66.
24. Soille, P., *Morphological Image Analysis: Principles and Applications*. 2003: Springer-Verlag New York, Inc. **391**.
25. Legland, D.; Arganda-Carreras, I. & Andrey, P., *MorphoLibJ: integrated library and plugins for mathematical morphology with ImageJ*, Bioinformatics (Oxford Univ Press), 2016. **32**(22): p. 3532–3534.
26. Meyer, F. and S. Beucher, *Morphological segmentation*. Journal of visual communication and image representation, 1990. **1**(1): p. 21–46.
27. Adams, R. and L. Bischof, *Seeded region growing*. Pattern Analysis and Machine Intelligence, IEEE Transactions on, 1994. **16**(6): p. 641–647.
28. Frise, E., *Level Sets*, in *Fiji*. http://imagej.net/Level_Sets.
29. Castle, M.K., J., *Rolling Ball Background Subtraction*, in *ImageJ and Fiji*. http://imagej.net/Rolling_Ball_Background_Subtraction.
30. Landini, G., *Auto Local Threshold*, in *ImageJ and Fiji*. http://imagej.net/Auto_Local_Threshold.
31. Kass, M., A. Witkin and D. Terzopoulos, *Snakes: Active contour models*. International journal of computer vision, 1988. **1**(4): p. 321–331.
32. Osher, S. and J.A. Sethian, *Fronts propagating with curvature-dependent speed: algorithms based on Hamilton-Jacobi formulations*. Journal of computational physics, 1988. **79**(1): p. 12–49.
33. Sliusarenko, O., et al., *High-throughput, subpixel precision analysis of bacterial morphogenesis and intracellular spatio-temporal dynamics*. Molecular Microbiology, 2011. **80**(3): p. 612–627.
34. Paintdakhi, A., et al., *Oufti: an integrated software package for high-accuracy, high-throughput quantitative microscopy analysis*. Molecular Microbiology, 2016. **99**(4): p. 767–777.
35. Delgado-Gonzalo, R., *E-Snake*, in *ImageJ*. http://bigwww.epfl.ch/algorithms/esnake/.
36. Delgado-Gonzalo, R., *Active cells*, in *Icy*. http://icy.bioimageanalysis.org/plugin/Active_Cells.
37. Delgado-Gonzalo, R., *Active Cells 3D*, in *Icy*. http://icy.bioimageanalysis.org/plugin/Active_Cells_3D.
38. Dufour, A., *Active contours*, in *Icy*.
39. Dufour, A., *3D Active Meshes*, in *Icy*. http://icy.bioimageanalysis.org/plugin/Active_Cells_3D.
40. Dufour, A., et al., *3-D Active Meshes: Fast Discrete Deformable Models for Cell Tracking in 3-D Time-Lapse Microscopy*. IEEE Transactions on Image Processing, 2011. **20**(7): p. 1925–1937.

41. Delgado-Gonzalo, R., N. Chenouard and M. Unser. *Fast parametric snakes for 3D microscopy*. in *Biomedical Imaging (ISBI), 2012 – 9th IEEE International Symposium on*. 2012. Ieee.

42. Delgado-Gonzalo, R., N. Chenouard and M. Unser, *Spline-Based Deforming Ellipsoids for Interactive 3D Bioimage Segmentation*. IEEE Transactions on Image Processing, 2013. **22**(10): p. 3926–3940.

43. Argandas-Carreras, I., et al., *Trainable Weka Segmentation*, in *Fiji*. http://imagej.net/Trainable_Weka_Segmentation.

44. Argandas-Carreras, I., S. Saalfeld and J. Schindelin, *SIOX: Simple Interactive Object Extraction*, in *Fiji*. http://imagej.net/SIOX.

45. Sommer, C., et al. *ilastik: Interactive learning and segmentation toolkit*. in *Biomedical Imaging: From Nano to Macro, 2011 IEEE International Symposium on*, 2011. IEEE.

46. Ouyang, W., *Rapid Learning*, in *Icy*. http://icy.bioimageanalysis.org/plugin/Rapid_Learning.

47. Hervé, N., *Texture Segmentation*, in *Icy*. http://www.herve.name/pmwiki.php/Main/TextureSegmentation.

48. Hervé, N., et al. *Statistical color texture descriptors for histological images analysis*. in *2011 IEEE International Symposium on Biomedical Imaging: From Nano to Macro*. 2011.

49. Perkin-Elmer, *inForm*. http://www.perkinelmer.com/catalog/category/id/inform-advanced-image-analysis-software.

4

Measuring Molecular Dynamics and Interactions by Förster Resonance Energy Transfer (FRET)

Aliaksandr Halavatyi and Stefan Terjung
European Molecular Biology Laboratory (EMBL), Heidelberg, Germany

4.1 FRET-BASED TECHNIQUES

Techniques based on the Förster resonance energy transfer (FRET) phenomenon are frequently applied to investigate molecular interactions between molecules of interest, measuring physiological parameters in living samples via sensors based on FRET and in specialised cases to measure distances on molecular scales using single molecules (molecular ruler).

Theodor Förster was the first who described the effect of resonance energy transfer [1]. FRET can occur if an exited fluorescent molecule (donor) is in very close proximity to another molecule (acceptor) (Figure 4.1). The energy transfer is a non-radiative process via dipole-dipole interaction which only occurs if three conditions are met [2]:

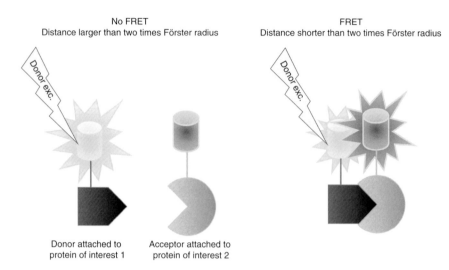

Figure 4.1   FRET phenomenon is used to measure association kinetics of biomolecules. Sample is illuminated in the donor excitation spectral range. If donor and acceptor molecules are far away from each other (left) emission is detected only in the donor channel. When the two fluorophores get closer (right), FRET can occur, resulting in decrease of donor emission and increase of acceptor emission.

- sufficient overlap between donor emission and acceptor absorption spectra (appropriate level of transferred energy)
- distance between donor and acceptor less then twice Förster radius R_0, which is defined as the distance at which 50% FRET efficiency occurs for a given donor-acceptor pair
- orientation of the fluorophore dipoles not perpendicular

The Förster radius of typical FRET pairs is in the range of 2–7 nm and therefore in the scale of biomolecular sizes, rendering FRET very attractive for investigating molecular interactions in biology.

In this chapter we focus on two FRET acquisition and processing techniques that do not require specialised hardware: ratiometric imaging and acceptor photobleaching. Reviews on FRET are available for more detailed information [3, 4].

4.1.1   Ratiometric Imaging

This method is the easiest way to image FRET samples, but it only works when donor and acceptor are parts of a single molecule, as is often the case for FRET-based sensor molecules. A large variety of biomolecular sensors is available, which are based on changes of intramolecular FRET due to interactions of the sensor with the molecule of interest [5].

There are several classes of sensors, which differ in the way the intramolecular FRET efficiency is changed [6]:

- conformational changes induced by binding of the biomolecule, e.g. yellow cameleon [7]
- conformational changes induced by (de-)activation due to enzyme activity like phosphorylation of the sensor
- cleavage of the sensor separating donor and acceptor, e.g. to detect activity of proteases [8]

For this method, donor and acceptor imaging is performed under donor excitation. Upon change of the intramolecular FRET efficiency, the ratio between donor-emission and acceptor-emission indicates the relative FRET changes (Figure 4.2). High FRET efficiency corresponds to relatively low donor emission and relatively high acceptor emission (Figure 4.1). Low FRET efficiency reverses this trend.

4.1.2 Acceptor Photobleaching

The energy transferred via FRET onto the acceptor results in reduced fluorescence emission of the donor since donor molecules return to ground state after transfer, without emitting a photon. This effect can be used to measure the FRET efficiency by acquiring a donor and acceptor image before and after bleaching a region of interest (ROI) in the acceptor channel. In the case of FRET, the donor emission in the ROI will increase and the difference between pre- and postbleach donor intensity directly refers to the relative number of donor molecules involved in FRET (Figure 4.3). Acceptor photobleaching is limited to analysing fixed samples, since movement between pre- and postbleach images results in strong artefacts.

4.1.3 Other FRET Measurement Techniques

4.1.3.1 Sensitised Emission

Another option to measure FRET is to detect the fluorescence of the acceptor under donor-excitation (sensitised emission). Unfortunately most of the time this sensitised emission signal is superimposed by other fluorescence contributions and the correction for these contributions via

(a)

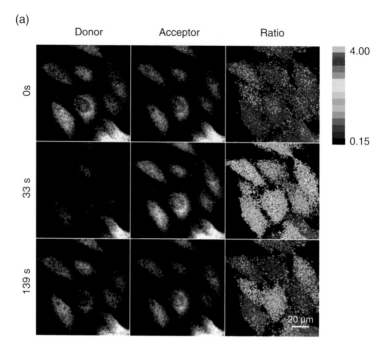

(b)

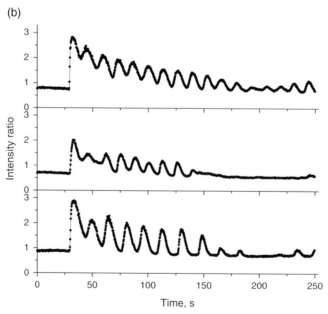

linear unmixing complicates this method. The strongest contributions to the donor-excited acceptor channel result from emission of the donor (bleed-through) and direct excitation of the acceptor (cross-excitation). An advantage of this method is that once all factors are measured for given imaging conditions, it can be used to visualise relative FRET differences in living samples.

4.1.3.2 Fluorescence Lifetime Imaging Microscopy (FLIM)

Each fluorophore has a characteristic average time it takes from absorbing an excitation photon until emission of a fluorescence photon. The fluorescence lifetime is defined as the time it takes until half of the excited molecules have emitted a photon. Energy transfer is an additional process of how donor molecules can return to the ground state. Consequently the lifetime of donor molecules is decreased if FRET occurs.

4.1.4 Alternative Methods to Measure Interactions

The FRET signal detected even at high FRET efficiencies is in most cases a relatively small contribution to the total fluorescence signal. To detect a very low number of interacting molecules over the dominating fluorescent background of non-interacting molecules is very difficult. As an alternative in such cases we mention two related techniques.

Figure 4.2 Measuring intracellular calcium levels using FRET-based sensors and ratiometric imaging.
HeLa cells transfected with the calcium sensor yellow cameleon YC3.6 [7], containing both donor (cyan FP) and acceptor (yellow FP) framing calcium sensing elements. Upon binding of calcium the conformation of the sensor changes, bringing both FPs closer together. It leads to an increased FRET efficiency and corresponding ratio increase. A time-lapse of 1000 time points at an interval of 250 ms was acquired. After 100 time points (25 s), ATP was added to the medium to stimulate calcium oscillations.
(a) Example images of the time-lapse at 0 s, 33 s and 139 s showing the donor and acceptor channel and the calculated ratio image using the 16-colour lookup table of Fiji (black pixels correspond to NaN values after thresholding). Note the relatively high noise level in the single images, which was tolerated on purpose to enable a high acquisition rate without too much photobleaching and phototoxicity. Even at this high noise level it is possible to visualise spatially distributed ratio changes.
(b) Quantifying the average ratio (e.g. cell by cell) allows comparing of the behaviour of different cells or compartments over time. Note that averaging ROIs of the rather noisy images results in smooth curves enabling detailed comparison of the cells, revealing their individual behaviours.

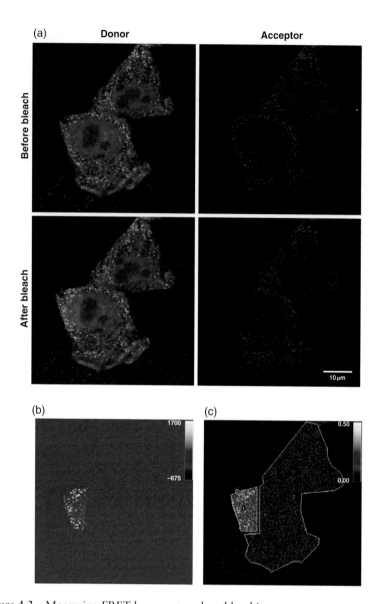

Figure 4.3 Measuring FRET by acceptor photobleaching.

(a) Hela cells transfected with a positive control linking GFP via linker to mCherry. Donor images are shown in green and acceptor images in red. The top and bottom rows display the snapshots before and after acceptor bleaching.

(b) Donor unquenching $(D_{post} - D_{pre})$ image using the fire lookup table of Fiji. Note the significant increase of donor fluorescence in the part of the cell where the acceptor was bleached. The values in the area covering the unbleached part of the cell fluctuate around zero. This is mainly caused by the noise in the raw data and sometimes by sample drift between prebleach and postbleach image acquisition.

(c) Apparent FRET efficiency $(D_{post} - D_{pre}/D_{post})$ in fire lookup table of Fiji. Black pixels correspond to NaN by thresholding D_{Post} before division. The average value in ROI 1 (where the acceptor was bleached) is 0.252 corresponding to 25% apparent FRET efficiency. The approximate error of the measurement can be estimated by averaging the part in which the acceptor has not been bleached (ROI 2: 0.016 corresponding to 0.016/0.252 = 6.3%).

4.1.4.1 Fluorescence Complementation

Fluorescent proteins like the green fluorescent protein (GFP) can be reconstituted from two (well-designed) non-fluorescent fragments, if they are in close proximity [9, 10]. This effect, which is often called bimolecular fluorescence complementation (BiFC), can be exploited to probe whether two molecules interact, by fusing one of the non-fluorescent fragments to each of the molecules of interest. Fluorescence will only be detected if both molecules are in direct contact. With appropriate controls this method can be a strong indication of interactions between the molecules of interest. A big advantage is that this method is sensitive for a low number of interactions, because the induced fluorescence can be detected above the dark background. The main disadvantage is that once the fluorescent molecule is formed, it is stable and stays fluorescent. Therefore it is not possible to measure dynamics with BiFC. Recently, tripartite complementation has been introduced as an improved version of BiFC [11]. In this method two GFP-fragments of 20 amino acids (β-strands 10 and 11) can be fused to the proteins of interest. If these two molecules interact, the combined β-strands 10 + 11 can complete the third component containing GFP β-strands 1–9 to a functional GFP molecule. Advantages of the tripartite complementation are higher solubility for the fusion proteins, smaller tags and reduced background due to lower self-association.

4.1.4.2 Dimerisation Dependent Fluorescent Proteins (ddFP)

Dimerisation dependent fluorescent proteins (ddFPs) can be used very similarly to BiFC. A clear increase of fluorescence is achieved upon heterodimerisation of an A-form exhibiting very low fluorescence as a monomer and a non-fluorescent B-form. A prerequisite to use such couples as biosensors is a low tendency for heterodimerisation. At present there are green, yellow and red versions of such A-forms described [12, 13]. The observation that the green and the red A-form both heterodimerise with the same non-fluorescent B-form opens up possibilities of designing new types of biomolecular sensors using the switch from one form to the other as a ratiometric readout [14]. A big advantage of ddFPs is the reversibility of the heterodimerisation enabling the readout of dynamic processes with this method.

4.2 EXPERIMENTAL DESIGN

A general and important prerequisite for quantitative experiments including FRET is to avoid saturated pixel values and the suppression of weak signals by too extensive offset settings.

Very often fluorescent proteins are used for FRET experiments. The optimal way to perform these kinds of experiments is to create stable cell lines with replacement of endogenous alleles by the FP-tagged version of the protein of interest, which is possible for example by CRISPR/Cas-based genome engineering [15]. Unfortunately genome engineering methods are still rather laborious and time consuming and a lot of experiments are performed using transient overexpression. Overexpression in general poses the risk of influencing the process of interest by flooding the cell with the expressed construct. In order to rule out effects due to overexpression, cells with low to medium expression levels need to be used and the relation between expression level and measured parameter has to be controlled.

Here we describe how to design, perform and analyse experiments using ratiometric imaging of FRET-based sensors and acceptor photobleaching. Both techniques can be performed using standard equipment like current confocal laser scanning microscopes, and analysis of the data is feasible using ImageJ (Fiji), which we explain in Section 4.4.2.

4.2.1 *Ratiometric Imaging of FRET-Based Sensors*

Since both donor and acceptor are on the same molecule, relative FRET changes can be measured by exciting the donor and detecting the emission of donor and acceptor in two separate channels. This can be set up as simultaneous acquisition using a confocal laser scanning microscope with two detectors, each channel adjusted to acquire one of the fluorophores efficiently with minimal bleed-through. It is also possible to use a wide-field or spinning disc confocal microscope for ratiometric imaging. Simultaneous acquisition can be achieved on such a system either by applying image splitting optics to project the two channels next to each other on the same camera chip or with a dual camera setup. A big advantage of wide-field and spinning disc confocal simultaneous acquisition is much higher frame rates compared to a point scanning confocal, and typically less photobleaching, but it comes at the cost of alignment of the channels with subpixel accuracy, otherwise the calculations to derive the ratio of the channels will lead to artefacts. Another option is sequential acquisition of donor and acceptor emission using a fast emission filter wheel. An advantage of sequential acquisition is that usually no image alignment is needed, but i) the channels have a slight time delay, ii) the excitation light dose increases and iii) acquisition time doubles.

This method is well suited to acquire time-lapse sequences to investigate dynamics. Quantitative measurements are possible with this technique using calibration. In case of calcium sensors e.g. yellow cameleon, calibration can be performed by first adding ionomycin in combination with a known calcium concentration in the medium to get the ratio at the high level (known concentration) and subsequently measuring the ratio at zero-level in presence of BAPTA-AM and EGTA.

4.2.2 Acceptor Photobleaching

Acceptor photobleaching experiments are frequently performed on confocal laser scanning microscopes, because the bleaching step is very easy to perform with the available bleaching functions. In general, all microscopes that are equipped to bleach an ROI are suited for this method. Firstly, a prebleach image set of the donor and acceptor channel is acquired, an ROI containing a part of the cell of interest including some background is bleached in the acceptor channel and finally the postbleach images of both channels is recorded. An important step of this protocol is the amount of bleaching. Complete bleaching of the acceptor in the ROI is needed to get the maximum unquenching of the donor. Only in this case can the real FRET efficiency be obtained, but bleaching of the donor by the bleach pulse needs to be avoided since this would decrease the calculated FRET efficiency.

A few controls are essential for acceptor photobleaching experiments, and all of them need to be acquired with identical imaging conditions as in the sample to be tested for FRET:

- donor only to check how much the donor channel is bleached by the acceptor bleach pulse and the acquisition settings
- acceptor only to control for photoconversion of the acceptor by the bleach pulse into donor-like fluorescent species – this photoconversion can occur to a different extent depending on the acceptor fluorophore used. Small contributions of a converted acceptor in the donor channel after bleaching can be corrected [16]
- negative sample containing both donor and acceptor, but without interaction causing FRET – in an optimal negative control the fluorophores need to be in the same environment or compartment as the sample to be tested for FRET. Especially if the FRET measurement is performed for proteins in small compartments like vesicles, it is important to control for FRET due to crowding. A convenient negative control to be used in experiments probing the cytoplasm is a tandem protein containing both donor and acceptor separated far enough by a large protein to prevent FRET, like mEGFP-MBP-mCherry [17]

- positive control with known high FRET efficiency to make sure there are no systematic errors in the experiment – often tandem constructs linking donor and acceptor with a short flexible linker are used as positive controls

A variation of the classic one-step acceptor photobleaching is repetitive acceptor photobleaching [18, 19]. Instead of a single strong bleach pulse a series of less intense bleach pulses is interleaved with acquisition of donor and acceptor channels. This is a particularly useful method if the donor is bleached partially by the acceptor bleach pulse.

4.3 FRET DATA ANALYSIS

FRET experiments are in most cases performed to find out whether, where and how much two proteins interact (acceptor photobleaching, sensitised emission, FLIM). The measured FRET efficiency is an indicator of where and how strong molecules interact, but only FLIM can report on the number of molecules, while the other methods report a combination of the number of interacting molecules convolved with the average distance and orientation. In the case of FRET-based sensors (ratiometric imaging) the FRET change is only a tool to report biological parameters like the concentration of biomolecules (e.g. Ca^{2+}-sensors) or activity (e.g. enzymes).

4.3.1 Ratiometric Imaging

Processing steps of ratiometric imaging data are relatively easy. But it is not sufficient to just divide one channel by the other, mainly because noise and background influence these calculations very strongly, especially in the low intensity areas. To compensate for these effects the raw data is usually first filtered to reduce the noise, either by a mean or a Gaussian blur filter. The filter kernel size has to be the same for both channels and is usually adapted to the noise level and the amount of desired detail; very often a 3×3 pixel kernel is a good choice. In the next step, background is subtracted from both channels and at least one channel is thresholded to set the background (parts without cells) to 'NaN' (= not a number) to avoid calculating values in this area. Subsequently the resulting images of the channels are divided to calculate the ratio. The resulting values only correspond to relative changes in FRET efficiency unless calibration has been performed to transform the ratio values into calibrated units.

4.3.2 *Acceptor Photobleaching*

To quantify the amount of unquenched donor, the donor image before the bleach pulse (D_{pre}) is subtracted from the postbleach donor image (D_{post}). The apparent FRET efficiency (E_A) of the donor is calculated by normalising the amount of unquenched donor to the total amount of donor:

$$E_A = \frac{D_{post} - D_{pre}}{D_{post}} = 1 - \frac{D_{pre}}{D_{post}} = E \cdot \alpha_D \qquad (4.1)$$

This equation specifies the calculation of the apparent FRET efficiency E_A with acceptor photobleaching data, where

- D_{pre} and D_{post} are the donor fluorescence intensities before and after the bleach pulse
- E is the FRET efficiency
- α_D is the relative amount of interacting donor.

If α_D is unknown, the true FRET efficiency E cannot be determined by acceptor photobleaching.

4.3.3 *Data Averaging and Statistical Analysis*

Comparison of FRET results between different experimental conditions requires analysis of the variability of quantified fluorescence signals and extracted parameters. In particular, estimated parameter values are reported with measures of their uncertainty, such as standard deviations or confidence intervals for mean values.

When measuring cellular response to a certain treatment over time (e.g. as shown in Figure 4.2a) we might intend to average FRET curves measured on different cells to characterise cell population. For asynchronous cellular response to the treatment (Figure 4.2b) such averaging can hide important kinetic information such as cellular oscillations of FRET efficiency.

For very noisy fluorescent signals, FRET efficiency (e.g. Equation 4.1) can be calculated on donor and acceptor signals averaged over certain ROIs (e.g. individual cells). Such fluorescence averaging neglects spatial specificity of donor and acceptor fluorescence. If possible, we advise performing pixel-based calculation of FRET efficiency (as shown in Figures 4.2 and 4.3) and, if needed, subsequent averaging of FRET efficiencies for larger ROIs.

4.4 COMPUTATIONAL ASPECTS OF DATA PROCESSING

4.4.1 *Software Tools*

Image processing steps are performed with either free or commercial image processing packages. Fiji is one of the most popular public domain packages [20] for image processing in biomedical sciences. It is based on ImageJ [21] and includes plugins and update functionality to keep ImageJ, Java and plugins on current and compatible status. Using the included Bio-Formats plugin, it is possible to open nearly all available microscopy data types directly, including metadata such as pixel size and timestamps.

4.4.2 *FRET Data Analysis with Fiji*

4.4.2.1 *Common Image Preparation Steps*

- Open the relevant data set by *File > Open* (or Ctrl-O).
- Choose *Split channels* in the Bio-Formats import options window.
- Smooth the data by selecting *Processing > Filters > Mean* or *Gaussian Blur*. Typically a radius of 1 (corresponding to a kernel of 3×3 pixels) is sufficient. It is recommended to select the *preview* option to get an idea of how the filter changes the image. Different filter settings can be compared to find out the optimal settings, but in the end all data needs to be processed with the same filter settings, and this has to be described in the Materials and Methods part of publications.
- Convert the data into 32-bit: *Image > Type > 32-bit*.
- To subtract the background, draw an ROI in an area of the image that does not contain cells and measure the average intensity by the menu commands *Analyze > Measure* (or Ctrl-M) for both channels. Select the whole image again by *Edit > Selection > Select All* (or Ctrl-A). Subtract the measured value by *Process > Math > Subtract* for each channel.
- Threshold at least one channel to convert the background to NaN (not a number) to exclude it from calculations by *Image > Adjust > Threshold*. Select the checkbox "Dark background", adjust the lower threshold (upper slider) to have the cells of interest thresholded (shown in red), click *Apply* and accept with the *Set background pixels to NaN* selected.

4.4.2.2 Ratiometric Data

- Follow steps described in Section 4.4.2.1.
- Divide one channel by the other: *Process > Image Calculator* and select The acceptor and the donor images as *Image 1* and *Image 2* with *Divide* as operation. Activate *Create new window* and *32-bit (float) result*.
- Quantification of FRET ratio over time (Figure 4.2b) can be done in the simplest case with the procedure, described in Section 5.4.1 of this book (see also Figure 5.4).

4.4.2.3 Acceptor Photobleaching

- Follow steps described above in Section 4.4.2.1.
- Subtract the prebleach donor image D_{pre} from the postbleach donor image D_{Post}: *Process > Image Calculator*; select D_{post} as *Image1* and D_{Pre} as *Image2* and *Subtract* as operation; check *Create new window* and *32-bit (float) result*. Rename the result with *Image > Rename*, e.g. to *Unquenched*.
- Divide the resulting image by D_{Post}: *Process > Image Calculator*; select the result of last step (e.g. *Unquenched*) as *Image1* and D_{Post} as *Image2* and *Divide* as operation; check *Create new window* and *32-bit (float) result*.
- Change lookup table (*LUT*) to false colour e.g. *fire*: *Image > Lookup Tables > Fire*.
- Adjust brightness: *Image > Adjust > Brightness/Constrast* (Ctrl-Shift-C) click on *Set* and set *minimum displayed value* to 0 and *maximum displayed value* to e.g. 0.5 to show apparent FRET efficiencies between 0 and 50 %.

4.5 CONCLUDING REMARKS

FRET can be well combined with Fluorescence Cross-Correlation Spectroscopy (FCCS), in which fluorescence fluctuations of two fluorophores are detected simultaneously. The cross-correlation of the two channels delivers the amounts of the two molecules which are diffusing together, indicating interaction of the two molecules or at least their integration in the same complex. While spectroscopy methods such as FCS and FCCS only measure molecular mobility in the focus volume, FRET reports distributions of interacting molecules. Combining FRET and FCCS allows the quantification of spatial distributions and concentrations of the interacting molecules.

As described in Section 4.1.3, more advanced FRET measurement methods, such as Sensitised Emission and Fluorescence Lifetime Imaging (FLIM), require either special optical equipment or additional data normalisation procedures. Issues of proper experiment installation and data processing are covered in dedicated reviews (e.g. [22] for FLIM).

4.6 REFERENCES

1. Förster, T., *Energy transport and fluorescence [in German]*. Naturwissenschaften 1946. **33**: p. 166–175.
2. Vogel, S.S., C. Thaler and S.V. Koushik, *Fanciful FRET*. Sci STKE, 2006. **2006**(331): p. re2.
3. Jares-Erijman, E.A. and T.M. Jovin, *FRET imaging*. Nat Biotech, 2003. **21**(11): p. 1387–1395.
4. Pietraszewska-Bogiel, A. and T.W.J. Gadella, *FRET microscopy: from principle to routine technology in cell biology*. Journal of Microscopy, 2011. **241**(2): p. 111–118.
5. Zhang, J., S. Mehta and C. Schultz, *Optical Probes in Biology*, vol. **6**. 2016: CRC Press.
6. Palmer, A.E. et al., *Design and application of genetically encoded biosensors*. Trends Biotechnol, 2011. **29**(3): p. 144–52.
7. Miyawaki, A. et al., *Fluorescent indicators for Ca2+ based on green fluorescent proteins and calmodulin*. Nature, 1997. **388**(6645): p. 882–7.
8. Hu, H.Y. et al., *FRET-based and other fluorescent proteinase probes*. Biotechnol J, 2014. **9**(2): p. 266–81.
9. Ghosh, I., A.D. Hamilton and L. Regan, *Antiparallel leucine zipper-directed protein reassembly: Application to the green fluorescent protein*. Journal of the American Chemical Society, 2000. **122**(23): p. 5658–9.
10. Kerppola, T.K., *Visualization of molecular interactions using bimolecular fluorescence complementation analysis: characteristics of protein fragment complementation*. Chem Soc Rev, 2009. **38**(10): p. 2876–86.
11. Cabantous, S. et al., *A New protein-protein interaction sensor based on tripartite split-GFP association*. Scientific Reports, 2013. **3**: p. 2854.
12. Alford, S.C. et al., *A fluorogenic red fluorescent protein heterodimer*. Chem Biol, 2012. **19**(3): p. 353–60.
13. Alford, S.C. et al., *Dimerization-dependent green and yellow fluorescent proteins*. ACS Synth Biol, 2012. **1**(12): p. 569–75.
14. Ding, Y. et al., *Ratiometric biosensors based on dimerization-dependent fluorescent protein exchange*. Nat Methods, 2015. **12**(3): p. 195–8.
15. Mahen, R. et al., *Comparative assessment of fluorescent transgene methods for quantitative imaging in human cells*. Mol Biol Cell, 2014. **25**(22): p. 3610–8.
16. Seitz, A. et al., *Quantifying the influence of yellow fluorescent protein photo-conversion on acceptor photobleaching-based fluorescence resonance energy transfer measurements*. J Biomed Opt, 2012. **17**(1): p. 011010.

17. Huet, S. et al., *Nuclear import and assembly of influenza A virus RNA polymerase studied in live cells by fluorescence cross-correlation spectroscopy.* J Virol, 2010. **84**(3): p. 1254–64.
18. Amiri, H., G. Schultz and M. Schaefer, *FRET-based analysis of TRPC subunit stoichiometry.* Cell Calcium, 2003. **33**(5–6): p. 463–70.
19. Van Munster, E.B. et al., *Fluorescence resonance energy transfer (FRET) measurement by gradual acceptor photobleaching.* J Microsc, 2005. **218**(Pt 3): p. 253–62.
20. Schindelin, J. et al., *Fiji: an open-source platform for biological-image analysis.* Nat Methods, 2012. **9**(7): p. 676–82.
21. Schneider, C.A., W.S. Rasband and K.W. Eliceiri, *NIH Image to ImageJ: 25 years of image analysis.* Nat Methods, 2012. **9**(7): p. 671–5.
22. Berezin M.Y. and S. Achilefu *Fluorescence lifetime measurements and biological imaging.* Chemical reviews. 2010;**110**(5):2641–2684. doi:10.1021/cr900343z.

5

FRAP and Other Photoperturbation Techniques

Aliaksandr Halavatyi and Stefan Terjung

European Molecular Biology Laboratory (EMBL), Heidelberg, Germany

5.1 PHOTOPERTURBATION TECHNIQUES IN CELL BIOLOGY

Measurement of spatiotemporal kinetics of biomolecules is essential for addressing the organisation of biological systems at different scales. Objects that are larger than the spatial resolution of the microscope can be segmented and tracked in order to measure their mobility (Chapter 7). Proteins and lipids, however, are much smaller than the resolution of light microscopes. Since current light microscopy techniques are not capable of directly resolving kinetics and interactions in the crowded cellular environment, several methods that enable indirect measurements of these parameters have been established.

Fluorescence recovery after photobleaching (FRAP) and related techniques are used for qualitative and quantitative studies of spatio-temporal kinetics of fluorescently labelled molecular species. The first protocols to measure molecular mobility with selective photobleaching

Standard and Super-Resolution Bioimaging Data Analysis: A Primer, First Edition.
Edited by Ann Wheeler and Ricardo Henriques.
© 2018 John Wiley & Sons Ltd. Published 2018 by John Wiley & Sons Ltd.

were developed in the 1970s [2]. Since the 1990s, when confocal laser scanning microscopes with hardware and software tools for photo-bleaching experiments became commercially available, the number of FRAP applications increased tremendously [3]. Genetic coupling of proteins with green fluorescent protein (GFP) chimeras opened perspectives for ectopic expression of fluorochrome-tagged proteins enabling much easier experiments with fluorescently labelled live samples as needed for FRAP. Since then, FRAP has been applied to study kinetics of various molecular machineries. [1, 4–10].

While qualitative FRAP analysis can be performed rather easily, accurate estimation of quantitative parameters from FRAP data is often challenging [11, 12]. FRAP results shouldn't be over-interpreted by applying complicated FRAP analysis protocols for which underlying assumptions do not reasonably approximate experimental setups. Application of simple exponential fit of the recovery curves with few free parameters is often preferable in practice over sophisticated FRAP fitting models trying to mimic the underlying reaction-diffusion complexity in detail. Estimation of individual reaction rates and other biophysical parameters requires additional measurements to restrict parameter space and to validate results of the quantitative FRAP analysis.

5.1.1 Scientific Principles Underpinning FRAP

In fluorescence microscopy spatial changes in location of molecules over time can be imaged by making a subpopulation of molecules optically distinguishable. This can be achieved by changing the ability of fluorophores to emit photons (bleaching, photoactivation) or by affecting spectral properties of emitted light (photoconversion). FRAP, in its simplest approach, uses the native property of all fluorochromes to bleach when exposed to excitation light. Fluorochromes can be bleached rapidly with high-intensity illumination in their absorption spectrum. In FRAP experiments (Figures 5.1 and 5.2) high light power is selectively applied to a region of interest (ROI) of the cell. In this case, bleaching is detected as loss of fluorescence intensity in the selected region. The bleached molecules may be either: (a) mobile, so they will move out of the bleached region and get replaced by a similar amount of non-bleached molecules under steady-state conditions; fluorescence brightness in this region recovers at the rate of molecular exchange between bleached and non-bleached regions, or (b) stably bound ("immobile" on the experimental timescale) and so will not be replaced by new, non-bleached fluorescent molecules.

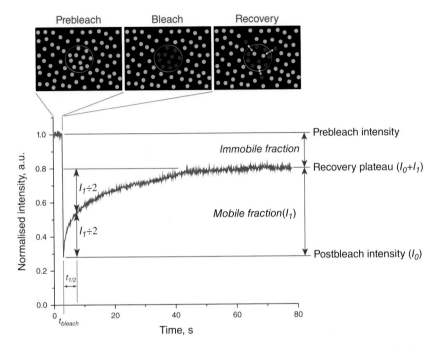

Figure 5.1 FRAP measures mobility of biomolecules. Steady-state spatial distribution of labelled molecules is detected from prebleach images. Bleaching of the selected ROI perturbs equilibrium of fluorescence distribution, which is restored over time due to replacement of the bleached fluorophores in the ROI with non-bleached fluorophores. FRAP recovery is quantified as an average or total intensity in the bleached ROI.

The initial fluorescence distribution between the bleach ROI and its neighbourhood will only be completely restored if all labelled molecules are mobile in the experimental timescale. In many biological processes this is not the case, and a fraction of the fluorescent molecules will be mobile (the mobile fraction), the rest will be stably bound (the "immobile" fraction, leading to incomplete fluorescence re-equilibration).

Time-lapse imaging of the cell after bleaching an ROI illustrates how fast molecules of interest exchange between the bleached region and its surrounding. Biological conclusions about mobility of the tagged molecules and the strength of their interactions can be deduced by collecting such movies under different conditions. Since the FRAP recovery rate is an indicator of molecular mobility, its quantification is of primary importance for quantitative processing of FRAP results. Although fluorescent recoveries are often visible to the eye, image analysis is required to compare between datasets. Quantification of recovery is

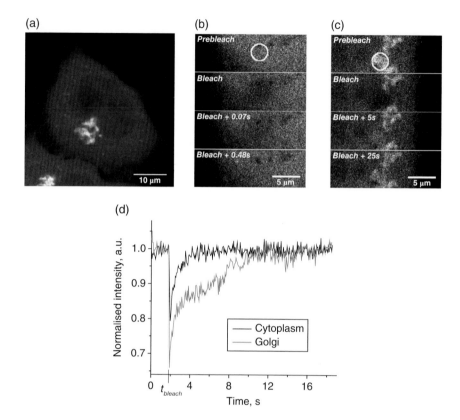

Figure 5.2   Use of FRAP to measure reaction-diffusion kinetics of molecules in living cells

(a) A HeLa Kyoto cell expressing Arf1-GFP. Fluorescence is distributed throughout the cytoplasm in agreement with the function of Arf1 in membrane trafficking. Expected continuous binding/unbinding of Arf1 at Golgi and its diffusional redistribution in the cytoplasm could not be measured because individual fluorophores are not identifiable with a confocal microscope. Molecules leaving a particular area are replaced by others at steady state.

(b) Extracted frames of the FRAP experiment with a circular bleach region in the cytoplasm.

(c) Extracted frames of the FRAP experiment with photobleaching Golgi membranes. Recovery is slower due to the transient arrest of molecular movements.

(d) Quantification of fluorescence intensity time-courses is required to compare recovery kinetics between different conditions. When curves are normalised to the prebleach fluorescence intensity in the bleached region, each individual value corresponds to the fraction of recovered fluorescence at a given time point enabling comparison of recoveries for different cells, cellular regions and conditions.

performed by measuring the average or sometimes the total fluorescent signal in the bleached region over time. The recovery kinetics can be inspected by plotting these numbers against corresponding time values (Figure 5.1). To compare the different measurements, all quantified curves are scaled to the same intensity range and normalised to correct for acquisition artefacts (Section 5.3.2). Recoveries that reach the plateau level faster correspond to more dynamic processes (Figure 5.2). The rate of FRAP recovery can be characterised by the half-time value (the value at which the recovery curve reaches its half-maximum, higher half-time values corresponding to slower recoveries).

The immobile fraction is often calculated as the relative difference between prebleach intensity and the recovery plateau. Quantitative processing of collected recovery curves can be used to determine:

- kinetics of biological processes (e.g. bleaching trafficking proteins)
- information about binding sites of proteins by making mutations in putative binding sites and comparing recovery kinetics with a wild-type form
- effect of drugs on binding characteristics of the molecule of interest
- comparison kinetics of the same protein in different cell compartments (Figure 5.2) to better understand reaction-diffusion kinetics [1].

Quantitative analysis of FRAP curves is performed by means of mathematical modelling. FRAP models can be either descriptive – e.g. depend on recovery halftimes and mobile fractions – or include mechanistic description of the underlying biophysical/biochemical processes in terms of diffusion/transport coefficients, binding constants and biochemical reaction rates. In Section 5.3.3 we discuss issues of FRAP model selection and their applications for estimating parameter values.

5.1.2 *Other Photoperturbation Techniques*

Influx and outflux of molecules relative to the bleached region can affect the shape of FRAP recovery. One way to uncouple parameters of interest is a combination of FRAP with related techniques, which make use of photo-switchable dyes, bleach fluorescence repetitively and measure fluorescence decays out of the bleached region to get complementary information about mobility of molecules within the cell. Here, we briefly present a commonly accepted classification of these methods and discuss respective biological questions to which they can be applied. It has to be noted, however, that combining several of these related techniques can be used to design an optimal photoperturbation experiment in a particular study.

5.1.2.1 Photoactivation (PA) and Inverse FRAP (IFRAP)

Photoactivatable proteins are converted from an initial non-fluorescent state to a fluorescent state, or from one emission spectrum to another (e.g. green to red) upon illumination in the appropriate spectral range (most often ~400 nm) [13]. Photoactivation (PA) of a limited cell region highlights tagged molecules of this region over dark background. Spatial redistribution of the activated molecules leads to the gradual decrease of fluorescence intensity in the photoactivated region. PA has the following advantages:

- allows tracking of the localisation of the converted molecules over time
- higher effective signal-to-noise ratio in comparison to FRAP as fluorescence signal over the dark background is quantified
- photoactivation of, for example, paGFP needs much less energy than a typical bleach pulse

Current drawbacks of available fluorescent proteins prevent frequent application of PA or photoconversion for measuring rapid reaction-diffusion kinetics. Some green-to-red photoconvertible fluorescent proteins (Dendra2, mEos2) need absorption of two photons sequentially for conversion to occur [14]. This results in rather slow conversion kinetics compared to photoactivation of paGFP or bleaching in the case of FRAP and can lead to an increase of green fluorescence instead of a switch. The higher the mobility of the investigated molecule, the more pronounced this effect is. Other photoconvertible fluorescent proteins like Kaede which do not show this effect occur in tetrameric form. This is very unfavourable for measuring dynamics of protein complexes. Without improved photoconvertible fluorescent proteins, currently the best alternative is to co-transfect the protein of interest fused to paGFP in combination with a second protein fused to a red fluorescent protein to highlight the structure of interest before photoactivation. This combination enables selection of cells with an appropriate transfection level, allows activation with a relatively low light dose and selection and tracking of the structure of interest using the red channel. The cost of this is that double transfection is needed, which slightly complicates the experimental protocol and might influence the cellular protein equilibrium.

Similar measurement of fluorescence redistribution in a certain region can be achieved with conventional fluorophores by bleaching the fluorescence of the entire cell except of the small ROI and measuring the signal trace in this ROI. This method is often named inverse FRAP (iFRAP) because of the reverse of the bleaching area compared to a classic FRAP experiment. Despite theoretical attractiveness, bleaching of such a big region requires an extended period of time and an increase of the needed light dose invoking additional cell stress and compromise quantitative data interpretation due to the movement of fluorophores during bleaching. The iFRAP method has been applied in the past mainly before the advent of paGFP to study rather slow processes like turnover of nuclear pore complex proteins [15].

5.1.2.2 Fluorescence Loss in Photobleaching (FLIP)

FRAP can be useful for studying the rates of molecular exchange between cellular compartments. Interpretation of FRAP in such experiments is complicated since bleached and non-bleached fluorophores can shuttle through intracellular membranes in both directions. Knowledge of both translocation rates is required to gain mechanistic understanding of this process; it is challenging to uncouple rates of molecule in- and outflux in FRAP measurements. For example, translocation of proteins through the nuclear membrane can be studied by bleaching fluorescence in the nucleus and following its recovery due to redistribution of fluorescent molecules from the cytoplasm into the nucleus. However, during the course of the experiment, proteins are still shuttling between the cytoplasm and the nucleus, and this can be a rate limiting factor for recovery. To measure one of these rates independently, fluorescence in one of these regions (e.g. the nucleus) can be continuously bleached while monitoring fluorescence decay in the adjacent region (e.g. the cytoplasm). Fluorophores are bleached whenever they are in the nucleus and do not contribute to cytoplasmic fluorescence afterwards even if they return to the cytoplasm. The measured decay curves are processed similarly to FRAP recoveries, estimating the rate of influx into the continuously bleached region and the immobile fraction of the tagged protein in the unbleached cell part [16].

5.1.2.3 Non-Steady-State FRAP, Fluorescence Localisation after Photobleaching (FLAP) and Spatial Intensity Redistribution Analysis

FRAP methods often assume that the spatial distribution of fluorescent molecules is in a steady state and that cellular organelles maintain their size, morphology and binding capacities throughout recovery curve acquisition. However, some cell structures and compartments undergo reorganisation on a similar timescale as turnover of molecules at them. To analyse fluorescence recoveries quantitatively in non-steady-state conditions, changes of total quantities of labelled proteins in the FRAP region have to be known [17]. Since some of the bleached molecules are lost through processes such as secretion or degradation, the total amount of tagged molecules (sum of fluorescent and bleached) can't be measured. FRAP can only be performed when molecules of interest are tagged with two spectrally distinguishable dyes where only one is bleached. Imaging of fluorescence kinetics in both bleached and non-bleached channels simultaneously provides data required for non-steady-state FRAP analysis. The results of two-colour time-lapse imaging can be processed in different ways:

- pixel-by-pixel subtraction of bleached channel images from reference images at each time point, also known as fluorescence localisation after photobleaching (FLAP), highlights regions of fast and slow molecular exchange [18]
- taking the ratio of signals in bleached and non-bleached channels, which allows normalisation of recoveries in non-steady-state conditions [19] or accounts for spatial organelle reorganisation [20]
- fluorescence traces in both channels can be used as independent input for a mathematical model that is designed to analyse non-steady-state photobleaching experiments [17].

For these experiments, it is important that the bleach and reference labels are sufficiently spectrally separated, so that there is no FRET between them. Although photoconvertible dyes might seem to be convenient for dual-colour kinetic analysis in non-steady-state cases, the nonlinear relationship of fluorescence intensity changes in two channels upon photoconversion renders such dyes currently inapplicable for quantitative studies of non-steady-state molecular kinetics. In the remainder of this chapter we focus on conventional FRAP, but most of the described modelling and data analysis principles are equally applicable for FLIP, PA etc.

5.2 FRAP EXPERIMENTS

The effectiveness of FRAP for drawing biological conclusions depends on appropriate implementation and proper use of the experimental protocols. FRAP is very sensitive to correct sample preparation, bleaching and image acquisition settings of the microscope for an accurate estimate of the kinetic parameters of underlying biological processes. In this section we highlight important details to be considered to achieve a solid experimental basis for FRAP analysis.

5.2.1 Selecting Fluorescent Tags

The ideal fluorophore should: (a) have no significant effect on the studied kinetic properties (mobility and interaction with other molecules) (b) be monomeric, (c) have high photostability so that acquisition photobleaching during the pre- and postbleach sequences is minimised, (d) be irreversibly bleached.

While acquisition photobleaching in general cannot be completely avoided, less than 20% acquisition photobleaching over the whole image sequence can be properly corrected by data normalisation.

Unfortunately, several fluorophores, especially FPs, reversibly transition into 'dark states', in which the molecules do not emit fluorescence. Often dark states are attributed to triplet states. The fraction of molecules in dark states depends on the illumination intensity and frequency. Therefore the dark state population is often increased by the FRAP bleach pulse. The return to the steady-state fluorescent population during postbleach imaging contributes to the measured recovery curve.

Whether a certain fluorescent protein or another fluorescent tag would be suitable for FRAP will depend on the target molecule under study, its intracellular localisation and other factors. It is highly recommended to refer to the dedicated literature [21] and include experiments for testing the effect of the tag on the rate and shape of recovery curves.

5.2.2 Optimisation of FRAP Experiments

Here we discuss the general guidelines which have to be considered in photobleaching experiments.

In a conventional FRAP experiment, bleaching has to be performed fast enough to minimise exchange of fluorescent and bleached molecules between the bleaching area and surrounding during the bleaching event.

Photobleaching has to be efficient enough to guarantee a sufficient dynamic range for the quantification of recovery acquisition. The optimum bleaching depth level depends on the used fluorophore and the studied biomolecule and therefore has to be determined experimentally.

The duration and frequency of time-lapse imaging has to be adjusted to obtain sufficient quantitative data to plot recovery kinetic curves without missing any data.

Acquisition of prebleach images is required to estimate the average fluorescence signal in the FRAP ROI before bleach. At least 5–10 prebleach images are usually acquired to be able to estimate fluorescence fluctuations and determine an average prebleach value. If the frame rate exceeds one frame per second, the number of prebleach images needs to be increased to at least 30 to reach a steady-state of the dark states at imaging conditions [22].

The number of postbleach images has to be sufficient to reach the fluorescence plateau of the recovery. Terminating acquisition before completion of fluorescence reequilibration might lead to biased estimates of the recovery parameters when slower processes are not taken into account.

Minimising acquisition photobleaching is done by balancing laser power, detector sensitivity, the acquisition frame rate and the number of acquired images.

It is of extreme importance to keep identical experimental settings (size and resolution of the imaged frame, acquisition frame rate, duration and power of bleaching, size and shape of the bleached region, etc.) throughout the experiment when acquiring technical and biological replicates. If some experimental settings are changed between acquisitions, the corresponding parameters have to be documented for each dataset individually and treated accordingly during analysis. For example, photobleaching with the differently sized ROIs can help to discriminate whether diffusion of the molecules is a limiting factor for the recovery (Section 5.3.3.2).

If the fluorescent signal varies considerably between cells (e.g. because of variable expression levels of the fluorescent protein) it might be necessary to adjust laser power or detector gain to guarantee proper dynamic range for the fluorescent signal and to avoid signal saturation. Because FRAP signals are usually normalised to the prebleach intensity levels, moderate changes of these settings should not have a drastic effect on the shapes of recovery curves. However, changing laser power would affect acquisition photobleaching, while adjusting detector gain has a strong effect on the signal-to-noise level. Therefore both changes can potentially affect the estimated values of the kinetic parameters and should be avoided as much as possible.

Often it is not possible to find the optimal settings in the first acquisition session. In such cases it is recommended to iterate between data acquisition and analysis until the desired quality of the data is achieved and these settings can be employed for the main experiment.

The recoveries have to be measured in several cells over three biological replicates. Five to ten recovery acquisitions per condition might be enough for a qualitative comparison of FRAP recoveries between multiple conditions in case of low biological variation and moderate noise levels. The number of FRAP datasets that need to be acquired for proper quantitative analysis depends on the variability of the shapes of individual recoveries and can vary between several dozen and several hundreds, as discussed in the Section 5.3.6.

5.2.3 *Storage of Experimental Data*

As for other microscopy experiments, images have to be well-documented and annotated properly as discussed in Chapter 10. Here we discuss which parameters are of critical importance for processing and interpretation of FRAP results.

If a manufacturer's FRAP addin has been used, it is advisable to use their proprietary format for data storage, as both the image data and metadata about the experimental conditions will be stored together in the same file. Important considerations are:

- frame rate of acquisition
- bleaching time and duration
- position, size and shape of the bleached region, as well as laser power and scan properties during the bleach
- time stamps for the acquired images
- the full series of images acquired before and after the bleaching as recommended practice for image analysis (Chapter 10)

Depending on the microscope brand, and the format of the saved data, some of these parameters are automatically stored as metadata for the acquired images or with intensity tables in text format (see Section 5.3.7). Bio-Formats library can generally be used to open this data in Fiji or other open-source software. We advise using the data of the trial experiment to discover which experimental settings can be retrieved from saved files and which of them have to be documented by the experimenter.

5.3 FRAP DATA ANALYSIS

As FRAP data processing involves multiple calculation steps, selection of computational tools can have a significant impact on the analysis speed and accuracy. This section is aimed to explain the general workflow for analysing FRAP recovery curves (Figure 5.3). Each of the highlighted steps can be performed in multiple ways, which are subsequently discussed.

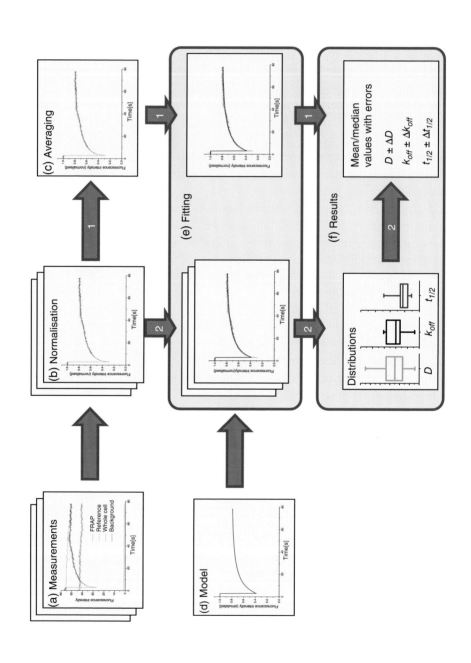

Figure 5.3 Workflow for quantitative analysis of FRAP recovery curves

Track 1 corresponds to estimations of recovery parameters from averaged recovery curves; in alternative track 2 each FRAP dataset is processed individually allowing to address biological heterogeneity for estimated kinetic and association parameters.

(a) Quantification of fluorescence intensity traces for FRAP, reference and background regions for multiple experimental measurements.

(b) Normalisation of recovery curves.

(c) Averaging normalised FRAP recoveries.

(d) Selection of mathematical model that is expected to match experimental data.

(e) Adjustment of model parameters to fit the data and evaluation of the fit quality.

(f) Calculating statistics: estimated parameters can be represented either by their mean/median values supplemented with error bars or by their distributions.

The choice of methods to employ and the configuration of the analysis pipeline depend on the addressed scientific question and the used microscope configuration. Here we cover only processing of the recovery curves measured in the bleached region. Spatial analysis of fluorescence redistribution after bleaching is performed similarly using appropriate mathematical models and computational analysis tools that mimic evolution of intensity profiles or fluorescence redistribution within the cell after bleaching [1, 23].

At the end of this section we provide guidelines for selecting the most suitable programs for the analysis of FRAP datasets. In Section 5.4 we describe several basic pipelines to process FRAP data using freeware programs.

5.3.1 Quantification of FRAP Intensities

In most protocols, the FRAP recovery signal is quantified by averaging the fluorescence intensities in the bleached ROI over the time course of the experiment. If protein turnover on a subcellular structure or an organelle is studied, the total intensity in the region covering the whole structure can be a more appropriate measure. This region is not influenced by slight alterations of the organelle shape and image blurring due to slight vertical fluctuations of the structure (acquisitions with too much defocusing of the structure have to be excluded from the analysis in any case). When these structures are smaller than the optical slice thickness, the fraction of fluorescence corresponding to the freely diffusing fluorescent species need to be subtracted from the total signal (see Section 5A.1 as an example).

Movement of bleached objects and organelles during acquisition will result in invalid recovery traces, if a fixed ROI is used for the readout. Consequently these traces have to be discarded from the analysis, if movement cannot be compensated by tracking the bleached structure or compartment in the FRAP image sequence. A fiducial marker such as an independent fluorescent marker targeting the same organelle or directly tracking the studied structure with the prerequisite of an incomplete bleach can assist faithful segmentation even in the first postbleach image.

In addition to the intensity trace in the FRAP ROI the average values of two reference regions are required for data normalisation (Figure 5.3a):

- Region 1 – a reference region inside the cell exhibiting a similar fluorescence level as the bleached region to correct for acquisition photobleaching and unspecific intensity changes during FRAP recovery
- Region 2: – a region without cells to quantify the background level.

The reference region must either (a) have the same geometry as the FRAP ROI or (b) cover the whole cell:

- FRAP ROI geometry – a region of the same size and shape as the bleach ROI can be used either in a neighbouring cell or in the same cell at an appropriate distance from the bleach area. The expected biochemical properties of FRAP and reference regions should be similar. When the object of study is a small cellular organelle, membrane or a structure like mitochondria, the reference region should be placed over a similar structure.
- Whole cell geometry – when a rather big region of a cell is bleached, a significant part of the cellular fluorescence is irreversibly lost. Normalising the recovery curve over the total cell intensity can often compensate for this loss and provide corrected estimates of the mobile and immobile fractions. For this normalisation a region covering the whole cell has to be quantified.

Preferentially, background and reference signals have to be measured on the same set of images that was used for quantification of the FRAP recovery. In some cases, it is not possible to acquire the FRAP dataset with appropriate reference and background regions. In such cases a workaround is to quantify these values from a separate image series acquired with exactly the same settings as the FRAP data except that the bleaching power is set to 0% and a background region is included. The latter is not optimal and is only an option for very stable microscopes for which fluctuations of the light sources and fluorescence detectors do not affect shapes of recovery curves.

5.3.2 Normalisation

Normalisation is performed before further processing of FRAP data to correct changes in the shapes of FRAP curves caused by the technical artefacts that lead to biased estimates of the recovery parameters. Among those are detector dark counts (photomultiplier for CLSMs or digital cameras), acquisition photobleaching and laser fluctuations.

The normalisation of recovery curves can be performed sequentially in the following order [22]:

1. Subtract the background fluorescence from the FRAP signal for each timepoint.
2. Calculate the average of the background-subtracted values for the prebleach time points. Then divide background-subtracted values for

all time points by this average. If a rapid intensity decrease is observed for the first few prebleach images, the corresponding values have to be excluded from averaging because it reflects equilibration of the dark state fraction for the applied time-lapse settings.

3. Perform steps 1 and 2 for the reference region.
4. Divide the corrected FRAP signal (2) by the corrected reference signal (3) to account for changes in fluorescence due to acquisition photobleaching.

These individual substeps are usually combined into a single equation to facilitate the calculation procedure. In case the FRAP signal is measured as average intensity in the bleached ROI, steps 1 and 2 can be combined to Equation (5.1) often called in literature the 'single normalisation method' because it only corrects for background level and scales the fluorescence brightness relative to the average prebleach intensity.

$$I_{norm}(t) = \frac{I_{frap}(t) - I_{back}(t)}{\left\langle I_{frap}(t) \right\rangle_{pre} - \left\langle I_{back}(t) \right\rangle_{pre}} \tag{5.1}$$

Including the correction using the reference region intensity (steps 1–4) results in Equation 5.2:

$$I_{norm}(t) = \frac{I_{frap}(t) - I_{back}(t)}{\left\langle I_{frap}(t) \right\rangle_{pre} - \left\langle I_{back}(t) \right\rangle_{pre}} \cdot \frac{\left\langle I_{ref}(t) \right\rangle_{pre} - \left\langle I_{back}(t) \right\rangle_{pre}}{I_{ref}(t) - I_{back}(t)} \tag{5.2}$$

This equation is often referred to as the 'double normalisation method' [24]. Notations in both equations are

- $I_{frap}(t)$ – intensity values in FRAP region
- $I_{ref}(t)$ – intensity values in reference region
- $I_{back}(t)$ – intensity values in background region
- $I_{norm}(t)$ – normalised intensity values
- $\left\langle ... \right\rangle_{pre}$ – averaging of the values for the prebleach time points.

When the FRAP and reference curves are noisy, calculation of the ratio by Equation (5.2) increases the noise of the normalised recoveries even further. In some cases smooth reference curves can be generated by fitting pre-processed reference data (step 3) with an exponential decay function. Fitted reference values can be used for data normalisation without noise amplification. Alternatively, acquisition photobleaching can be included in the FRAP fitting model with the fixed parameters estimated from the fit of the reference data [25].

Some protocols also suggest additionally scaling the dynamic range between 0 and 1 to remove the effect of incomplete bleaching from the recovery. The corresponding calculation formula is

$$I_{full\_norm}(t) = \frac{I_{norm}(t) - I_{norm}(t_{bleach})}{\left\langle I_{norm}(t) \right\rangle_{pre} - I_{norm}(t_{bleach})} \tag{5.3}$$

where t_{bleach} is the bleaching time; $I_{norm}(t)$ is the normalised intensity trace (result of calculations with Equation 5.1 or 5.2). In practice, the normalised intensity value at first postbleach time point is used as an approximation of $I_{norm}(t_{bleach})$, which could be a rather poor estimate of the bleaching depth (Note 1). Therefore we generally do not advise this additional scaling, unless the selected fitting equation implies 100% bleaching.

5.3.3 *In Silico Modelling of FRAP Data*

A mathematical model for FRAP analysis is a computational procedure that can mimic FRAP recovery for the expected process and a selected set of its parameters. It can be implemented as analytical equation(s) for calculating intensity values at each time point or as a computational algorithm that simulates exchange of fluorescent and bleached molecules between bleached ROI and its surrounding (Note 2). Certain assumptions about the process, the cell geometry and experimental settings are needed to formulate each FRAP model. Correct interpretation of a quantitative FRAP analysis strongly depends on an adequate match of these assumptions and experimental settings.

Plenty of mathematical models have been developed in past decades, which are tailored to the variety of studied phenomena and experimental designs. In this section, we describe the most important classes of FRAP models and briefly discuss distinctive features of data analysis and interpretation for each class.

5.3.3.1 *Descriptive FRAP Models*

The simplest recovery analysis can be performed with equations that parametrise recovery curves without assuming any particular kind of biophysical process. For many experiments FRAP curves can be well approximated by single exponential recovery:

$$FRAP(t) = I_0 + I_1 \cdot \left(1 - e^{-\frac{t - t_{bleach}}{\tau}} \right) \qquad (5.4)$$

where I_0, I_1 and τ are parameters defining the shape of the FRAP curve (Figure 5.1). I_0 is the normalised intensity just after bleach and I_1 is the dynamic range of recovery. The immobile fraction can be calculated from these values as $(1 - I_0 - I_1)/(1 - I_0)$. The rate of recovery depends in this equation only on the parameter τ from which the halftime of recovery can be derived: $t_{1/2} = \tau \cdot \ln 2$.

Some recovery curves exhibit a multiphase behaviour and are not well approximated by Equation 5.4 (evaluation of fit quality is discussed in Section 5.3.5). To better fit such data the previous equation can be extended with an additional term to biexponential equation:

$$FRAP(t) = I_0 + I_1 \cdot \left(1 - e^{-\frac{t - t_{bleach}}{\tau_1}} \right) + I_2 \cdot \left(1 - e^{-\frac{t - t_{bleach}}{\tau_2}} \right) \qquad (5.5)$$

Parameters of this equation have a similar meaning as in Equation 5.4: I_1 and I_2 are dynamic ranges for the first and second recovery components accordingly; τ_1 and τ_2 are characteristic times for those components.

Multiphase curves often occur in practice when recoveries are governed by two or more processes, for example binding and diffusion. This explains why the results of multiexponential fits are often misinterpreted by assuming that parameters I_1, τ_1 solely define the first process and I_2, τ_2 define the second process. The interdependence between exponential recovery components is very likely when rates of two processes do not differ a lot.

If necessary, the halftime of the full recovery can be estimated after multiexponential fitting by substituting estimated parameters to Equation (5.5), assigning its right side to 0.5 and solving for t.

The main disadvantage of analysing FRAP data with exponential equations or similar descriptive models is the lack of comparability between the estimated parameters and the results of other experiments. Extracted values have no direct biophysical or biochemical meaning and are only comparable for the data acquired on the same microscope with identical experimental settings.

5.3.3.2 Diffusion and Transport Models

FRAP was initially developed for measuring molecular mobility. Proper experimental design and quantitative analysis of recovery data can provide parameters for normal and anomalous diffusion or for directed flow.

The rate of FRAP recovery depends upon the time that the molecules – which are fluorescently labelled – need to move through the bleached region. Therefore, both size and shape of the bleached ROIs influence recovery. Bigger bleached regions result in slower recoveries. Each bleaching geometry and each kind of molecular movement needs to be analysed by the appropriate mathematical model. Strip-line bleaching ROIs (one-dimensional lateral diffusion) or circular ROIs (two-dimensional lateral diffusion with radial symmetry) are often used to simplify the modelling procedure.

Many molecules distribute through the cell body via Brownian motion, which is expressed mathematically by diffusion differential equations. FRAP models solve this equation to determine what percentage of the bleached molecules are replaced by their bright analogs at each recovery time point.

The first FRAP diffusion models assumed that bleaching occurs only within the bleached spot, and all molecules in the vicinity of the FRAP ROI remain unaffected by the bleach pulse [26]. More recent experimental and theoretical studies [11] showed that such an assumption often leads to significantly biased estimates of diffusion coefficients. Models and protocols for more accurate diffusion analysis also account for the following experimental factors:

1. Effective bleaching area is often larger than defined by the experimenter. First, the point spread function (PSF) of the laser beam causes bleaching of fluorophores in the vicinity of the FRAP ROI. Second, fast diffusion of molecules during the bleaching time increases effective bleaching size. Bleaching duration and PSF are included in some FRAP models. Alternatively, effective bleaching size can be estimated from acquired images and incorporated into calculations of the diffusion coefficient.
2. Bleaching efficiency is inhomogenous in the axial direction for objective lenses with high NA values. If the ROI is relatively deep inside the specimen (e.g. bleaching nucleus) simulation of 3D diffusion accounting for both lateral and axial bleaching inhomogeneity is required for accurate data analysis.

The diffusion coefficient can also be recalculated from the recovery halftime. Simplistic FRAP models for isotropic unlimited diffusion and circular regions [2, 26] predict that the diffusion coefficient can be estimated as

$$D \approx 0.22 \cdot \frac{r^2}{t_{1/2}} \tag{5.6}$$

where r is the radius of the bleached region.

As discussed above, for more accurate estimates, the effective bleaching radius has to be taken into account. In this case, Equation 5.6 transforms into [27]:

$$D \approx \frac{r_n^2 + r_e^2}{8t_{1/2}} \tag{5.7}$$

where r_n and r_e are the radii of nominal (defined by the experimenter) and effective (estimated from postbleach images) bleaching radii.

The relationship between diffusion coefficient and recovery halftime can be used to test whether FRAP data can be approximated by isotropic diffusion. Halftimes of FRAP recoveries are measured for several bleaching radii and plotted against the numerator term of Equations 5.6 or 5.7. Direct proportionality on this plot would indicate normal diffusion; deviations from direct proportionality are signs for restricted or anomalous diffusion. If $t_{1/2}$ values do not depend on the size of the bleached region, the recovery rate is not limited by the rate of molecular mobility per se, but by transient interactions of the labelled molecules with other cellular species.

5.3.3.3 FRAP Models for the Reaction-Dominant Regime

Biochemical interactions of labelled molecules prevent fluorescence reequilibration after bleaching governed by diffusion. The simplest reaction relevant for many biophysical processes corresponds to the reversible binding of freely diffusing molecules (F) to immobilised binding sites (S) represented by Equation 5.8:

$$F + S \longleftrightarrow C \tag{5.8}$$

where forward binding reactions occur with rate k_{on} and backward unbinding reactions occur with rate k_{off}. Transient complexes are represented by C.

FRAP recovery for a reaction scheme (5.8) can be represented as shown in Equation 5.9, with the assumption that the following prerequisites are fulfilled:

- the bleached ROI is very small in comparison with cell size
- binding sites are equally distributed in space
- bleaching is complete
- diffusion is very fast in comparison with binding and unbinding rates

$$FRAP(t) = 1 - c_{eq} \cdot e^{-k_{off} \cdot (t - t_{bleach})} \qquad (5.9)$$

where c_{eq} is the steady-state fraction of transiently bound molecules, defined as $c_{eq} = k_{on}^* / \left(k_{on}^* + k_{off} \right)$. Here k_{on}^* is the pseudo-binding rate constant equal to k_{on} multiplied by the concentration of available binding sites.

Equation (5.9) is applicable for analysing FRAP curves only when all underlying assumptions are met. Nevertheless, analysis of this equation provides a qualitative understanding of FRAP kinetics in reaction-dominant regimes:

- The dynamic range of normalised FRAP recovery is equal to the c_{eq} under the assumption of 100% bleaching. Unbiased estimation of c_{eq} is possible only when the actual bleaching depth is measured (e.g. with fixed cells) and included either to the fitting model or to the normalisation procedure.
- By analogy with Equation (5.4), the shape of the FRAP curve is solely defined by an exponential term. Consequently, only unbinding rate k_{off} (and not k_{on}) defines the rate of FRAP recovery. It holds true only in the steady-state regime with small bleaching ROIs, otherwise both k_{on} and k_{off} influence the FRAP recovery rate.
- The binding rate cannot be uncoupled from the number of available binding sites, only their product (k_{on}^* value) can be estimated. A higher percentage of binding sites will be occupied when the expression levels of labelled molecules are increased leading to lower estimated k_{on}^* values. Therefore only cells with similar expression levels should be selected to analyse the binding kinetics.

5.3.3.4 Reaction-Diffusion FRAP Models

The uncoupling of biochemical reactions from diffusion in FRAP data described above is not always possible. Models considering one or more interacting diffusion component are more problematic for quantitative analysis because they depend on many parameters.

Multiple modelling studies performed analytical and quantitative analysis for different combinations of diffusion and reaction parameters. In particular, the combination of isotropic diffusion with transient binding of diffusing molecules to the immobilised substrate has been intensively studied. It has been shown, that D, k_{on}^{*} and k_{off} values can be simultaneously determined from FRAP curves only in some parts of parameter space. Interestingly, in adjacent regions of parameter space FRAP curves for reaction-diffusion kinetics can be well approximated by diffusion-only governed recovery with reduced diffusion coefficient. Such FRAP regime is often called effective diffusion [28].

5.3.4 Fitting Recovery Curves

To compare a selected FRAP model with experimental data, it is necessary to find a set of model parameter values for which simulated results give the best approximation of the experimental data. A search for optimal parameters can be performed by running the simulation several times with different parameter sets and selecting those that provide the best fit. Because simultaneous optimisation of several recovery parameters is almost impossible to perform manually, it is normally executed using computational fitting procedures. In these procedures the difference between simulated and experimental data is defined by a numerical correspondence criterion. Smaller values correspond to a better model fit of the data. Although definitions of this criterion might vary between fitting algorithms, many of them are based on the sum of squared differences between theoretical and experimental data points. Functions for calculating this criterion are an integral part of the computational tools for FRAP data fitting and do not have to be defined by the experimenter.

The correspondence criterion is iteratively calculated for each tested parameter set. Parameters corresponding to the currently smallest criterion value are stored in memory.

The considerations below have to be taken into account to avoid misinterpretation of fitting results:

1. Good data fitting does not necessarily guarantee that the selected model is correct. Independent experimental tests are required to validate the selected model.

2. Many fitting algorithms start at a defined initial point in the parameter space and attempt to continuously improve fitting results by searching for neighbourhood points that provide a better fit. Such optimisation is terminated when the algorithm fails to further improve the correspondence criterion. Premature fitting termination might occur when initial parameters are far from their unknown optimal values especially for models with many parameters. Multiple runs of fitting with different initial parameters help to perform optimisation over the whole parameter space.

3. Several substantially different parameter combinations might give equally good fitting results for complex models. Such uncertainty can often be avoided by determining some parameters with independent measurements and keeping these values fixed during the fitting procedure. For example, the diffusion coefficient can be measured independently before fitting FRAP results with the model includes both binding and diffusion (Note 3).

4. Faster and more appropriate fitting results are achieved when possible ranges for optimised parameters are specified. Upper and lower limits for the rates of protein diffusion in cytoplasm, in the nucleus and on membranes can be estimated for a particular biomolecule [29]. Ranges for binding/unbinding rates of some processes can be extracted from biochemical or thermodynamic studies. The range for reliable estimates of characteristic recovery times (e.g. $t_{1/2}$ values or characteristic diffusion times) is limited by several time steps between adjacent data points and by half of the recovery acquisition time interval. If the rate of the studied process is found to be out of this range, experimental settings have to be reconsidered.

Immobile, free and bound fractions should be theoretically between 0 and 1 a slight increase of this range in both directions could give a better fit by accounting for experimental noise that can affect the first postbleach points and the fluorescence plateau level.

5.3.5 Evaluating the Quality of FRAP Data and Analysis Results

Even when all possible precautions during FRAP acquisition and processing are taken, evaluation of the data quality is needed to avoid erroneous conclusions.

1. Careful examination of the acquired image series is essential. Cells undergoing significant contraction or reorganisation during FRAP

acquisition are likely to be damaged by the light and have to be excluded from the analysis.

2. Recovery curves acquired during steady-state conditions are expected to reach plateau level at or below prebleach intensity levels. Recovery curves that grow far beyond this limit cannot be analysed with steady-state models.

3. Rates of recovery that have small relative difference between first and last acquired postbleach data points cannot be estimated with high precision (e.g. for high immobile fractions) therefore care should be taken when such data is to be quantitatively interpreted.

4. Inspection of the overlay between experimental and simulated recoveries is necessary, but not sufficient for accepting fitting results. Plot of residuals between two datasets is an additional control for fitting quality. Good fit residuals have to be randomly distributed around the time axis. Any kind of trend (e.g. wave shape) in the residuals plot indicates systematic deviation of the fitted curve from the data.

Because absolute values of the correspondence criterion are affected by experimental noise, they do not provide a measure of data fitting quality for different FRAP datasets. However, when normalised by the number of degrees of freedom (which is the number of fitted data points minus the number of model parameters) these values allow comparison of different model fits for the same dataset. For example, we can test whether a biexponential recovery provides a significantly better fit then the single-exponential.

5.3.6 Data Averaging and Statistical Analysis

In the case of FRAP, multiple approaches for statistical data evaluation are possible (Figure 5.3f):

1. Average curves are calculated from normalised experimental recoveries and then fitted by a mathematical model. Best fitted parameter values for the averaged recoveries are considered as representative parameter values. Weighted differences between data points and best fitted simulated recovery are converted to the confidence intervals for parameters with special statistical procedures.

   Although averaging of FRAP curves before fitting is helpful to reduce their noise, and is appropriate for their visual comparison, it can lead to significant inaccuracies in quantitative FRAP data analysis. First, high biological variability of the individual recovery rates can result in a non-representative shape of the averaged curve. For example, the average curve for single-exponential recoveries with broadly distributed halftimes would be well fitted only by a multiexponential equation. Second, estimation of parameter errors would only account for experimental noise and not for the differences in rates of individual recoveries leading to identification of false-positive differences between experimental conditions.

2. To perform unbiased estimation of parameters, each recovery curve has to be fitted independently. The quality of each fit is then evaluated to determine whether the selected mathematical model is representative for a particular dataset. Only accepted datasets are subsequently used to calculate the statistics of model parameters. When 10–30 recoveries are successfully processed, the average values with the corresponding standard deviations can be reported. Alternatively, by acquiring and processing a higher number of recoveries (typically 100 or more) one can analyse distributions of measured parameters. Comparison of these distributions with non-parametric statistical methods often helps to overcome misinterpretation of average values when experimentally measured kinetic parameters do not undergo normal distribution. However, a large number of representative datasets for multiple experimental conditions is hardly achievable with manually conducted FRAP experiments. High-throughput FRAP data can be collected using automated feedback microscopy technique [30, 31].

5.3.7 Software for FRAP Data Processing

Simple qualitative and quantitative analysis of already quantified FRAP curves can be performed manually using any spreadsheet software (e.g. MS Excel or Apache Open Office Calc), GUI-based statistical packages or custom scripts for scientific packages (R, MATLAB, Fiji/ImageJ). Several specialised tools for FRAP analysis have been designed to facilitate processing of big data collections and to provide unbiased estimates of underlying kinetic parameters (Table 5.1). In these tools, experimental recovery data is approximated with either analytical models or stochastic simulations (Section 5.3.3 and Note 2). Because there is no universal analysis pipeline to perform and to analyse FRAP experiments, none of these programs can be recommended as a general tool for FRAP data processing.

When interpretation of FRAP results and estimation of kinetic parameters are based on the recovery curves, quantification of fluorescence

Table 5.1 Software tools for FRAP data processing.

Tool/module name	Dependencies	Webpage	Main options
		Processing FRAP recovery curves	
FrapCalc	Macro for IgorPro (WaveMetrics)	https://github.com/cmci/FrapCalc	• Processing individual FRAP datasets • Data visualisation • Normalisation • Fitting with exponential and diffusion equations • Batch processing
easyFRAP [32]	standalone (Windows and Mac) or MATLAB script	http://ccl.med.upatras.gr/index.php?id=easyfrap	• Data visualisation • Normalisation • Fitting with exponential curves • Estimates of recovery halftimes and mobile fractions • Batch data processing • Calculating and processing mean recoveries
FRAPAnalyser	standalone (Windows)	https://github.com/ssgpers/FRAPAnalyser	• Data visualisation • Normalisation • Modelling and fitting recovery curves (exponential, binding, diffusion, binding + diffusion) • Estimates of recovery parameters (depending on the selected fitting model) • Batch data processing • Calculating and processing average recoveries

Stochastic simulations of FRAP data in constrained cell environments

Virtual FRAP (Virtual cell)	Part of the Virtual Cell package	http://wiki.vcell.uchc.edu/twiki5/bin/view/VCell/VFRAP http://vcell.org/	• Spatiotemporal simulation of fluorescence redistribution • Using imaging data to estimate kinetic parameters

All-in-one (quantification of intensities, modelling and fitting)

FRAP Calculator	ImageJ/Fiji macro	https://www.med.unc.edu/microscopy/resources/imagej-plugins-and-macros/frap-calculator-macro	• Quantification and normalisation of recovery curves • Noise correction • Estimation of mobile fractions and recovery halftimes
simFRAP	ImageJ/Fiji plugin	http://rsb.info.nih.gov/ij/plugins/sim-frap/index.html	• Simulation of 2D diffusion as random walk • Quantification of FRAP recoveries on experimental and simulated image series • Estimation of diffusion coefficients
Tropical [33]	standalone (Windows)	http://ibios.dkfz.de/tbi/index.php/software/further-software/101-tropical-1-0	• 2D spatiotemporal reaction-diffusion modelling of fluorescence redistribution after photobleaching (ODE based) • Estimation of model parameters
FRAP Monte Carlo simulation environment [23]	Java	(available by request from the authors)	• 3D spatiotemporal particle-based simulation with the possibility to specify compartments and their properties • Simulation of bleaching in 3D with defined laser profile • Estimation of model parameters

signals is often uncoupled from further data analysis steps, i.e. two independent programs are used to extract fluorescence intensities from the images and to process these numbers. Quantification results are typically stored as numeric tables where each time point is represented by a single row. One column normally contains imaging times relative to the start of FRAP acquisition, while others store intensity values for each measured ROI. If no image pre-processing or tracking of bleached structures are required, mean ROI intensities over time can be measured with most of the microscope control software immediately after FRAP acquisition. Alternatively, images are opened with image processing software (e.g. Fiji) to get time courses of intensities for FRAP, reference and background ROIs, together with required supplementary measurements, such as nominal and effective radii of the bleached region. If possible, images have to be imported into the software together with their metadata of the FRAP experiment, in particular position, size and shape of the bleached region, start and duration of the bleach and time stamps.

Exported table files with time and intensity values can then be processed manually with a spreadsheet software, or with a custom script or by one of the packages for analysis of FRAP curves. These packages provide a GUI to control the steps of the FRAP data processing. Three packages listed in Table 5.1 essentially follow the data processing approach in Figure 5.3 to convert intensity tables into the kinetic parameters characterising the process under study. The main differences between those packages are the accepted input file formats and implemented FRAP models for fitting.

Processing FRAP measurements 'one-by-one' is very time-consuming when working with large data collections. Therefore several programs (e.g. easyFRAP and FRAPAnalyser) are equipped with options for batch data processing. In these tools, the whole analysis pipelines or individual steps (e.g. normalisation or fitting) can be performed for a number of FRAP measurements at once. The fitted parameter values for each curve are exported in an output table for further statistical analysis. Simultaneous loading of all FRAP datasets into these programs enables calculating average recovery curves and fitting them with implemented kinetic FRAP equations.

Tools for spatiotemporal FRAP analysis simulate redistribution of bleached and non-bleached molecules in 2D or 3D. Some of these tools take into account cellular compartmentalisation, which is restricting diffusion and limiting biochemical reactions. Collected time-lapse images and coordinates of cell boundaries have to be loaded into these tools as input data. Parameter fitting is implemented either by extracting recovery curves from both experimental and simulated time-lapse images or by calculating similarity factors between those images.

Except the programs listed in Table 5.1 modelling tools for cell biology (e.g. COPASI, Virtual Cell, Berkley Madonna) are often used to create models for FRAP experiments and to perform data fitting.

5.4 PROCEDURES FOR QUANTITATIVE FRAP ANALYSIS WITH FREEWARE SOFTWARE TOOLS

5.4.1 *Quantification of Intensity Traces with Fiji*

Analysis of FRAP or, time-lapse imaging data involves measurement of fluorescence intensities in several cell regions for all experimental time points. Here we provide a pipeline to quantify intensity traces in fixed ROIs with standard Fiji options:

- Open the acquired time-lapse images in Fiji. In most microscopes prebleach and postbleach time-lapse images are stored in a single file. Use Bio-Formats plugin (*Plugins > Bio-Formats > Bio-Formats Importer*) to open the raw image files including the metadata. In the appearing options menu select the *Display ROIs* option.
- Create a list of ROIs for which intensity traces have to be quantified using the *ROI Manager* window (for FRAP it usually will be bleaching, reference and background ROIs). If some ROIs were extracted from the metadata, they will appear in the left part of the *ROI Manager* window. To create or change the set of ROIs in the *ROI Manager* manually:
  - Open the *ROI Manager* window in case it was not opened before (*Analyse > Tools > ROI Manager...*).
  - Draw single ROI with one of the selection tools and then press the *Add* button in the *ROI Manager* window to add the ROI to the list (Figure 5.4a,b).
  - Click on any ROI in the list to check its location and shape. The outline of the ROI will appear on the active image.
  - To delete an ROI, select it in the *ROI Manager* and click the *Delete* button.
- Specify which measurements have to be performed (*Analyse > Set Measurements*). For example, in conventional FRAP analysis only *Mean gray value* option has to be selected (Figure 5.4c).
- In the *ROI Manager* window, select all ROIs for which measurements have to be performed (e.g. while holding the *Shift* key click on the first

and the last to mark a range; alternatively hold the *Ctrl* key and click on all desired ROIs sequentially).

- Run quantification by selecting *More > Multi-Measure* in the *ROI Manager*. In the appearing dialog keep the options as specified in Figure 5.4d. A new *Results* window with quantification output should appear (Figure 5.4e).
- Save the content of the *Results* window as an XLS file (*File > Save as…*).

5.4.2 *Processing FRAP Recovery Curves with FRAPAnalyser*

This pipeline highlights the processing of FRAP curves in FRAPAnalyser, software that allows specifying the steps of a FRAP processing protocol through a user-friendly interface. The complete user manual can be downloaded, together with the software (Table 5.1).

- Prepare input data files with any spreadsheet software (e.g. MS Excel). Each FRAP dataset has to be represented by a table stored as a tab-delimited text file. The first row in this table contains the column names; all other rows contain only numerical values. The first column contains time values in seconds, the other columns contain intensity values for quantified ROIs (e.g. bleached, reference and background ROIs).
- Start the FRAPAnalyser software.
- Load the prepared text files (*Data > Load experiment*; select one or several files in the *Input Files* dialog window).
- In the automatically appearing *Set ROIs* window select data columns corresponding to the measured FRAP, reference and background signals.
- Specify the first postbleach time point in the appearing *Set time points* window.
- If needed, set actual bleach time for each dataset (*Data > Set Bleach Time*).
- Examine structure and content of the *Table* panel. You will see imported tables with additional columns reserved for analysis results.
- In the *Show on plot* drop-down list (*Settings* panel) select *All measured intensities* to visualise measured intensity time-courses.
- Normalise loaded datasets (*Normalisation > [One of normalisation methods]*).
- In the *Show on plot* drop-down list (*Settings* panel) select *Normalised curves* to plot the result of normalisation.

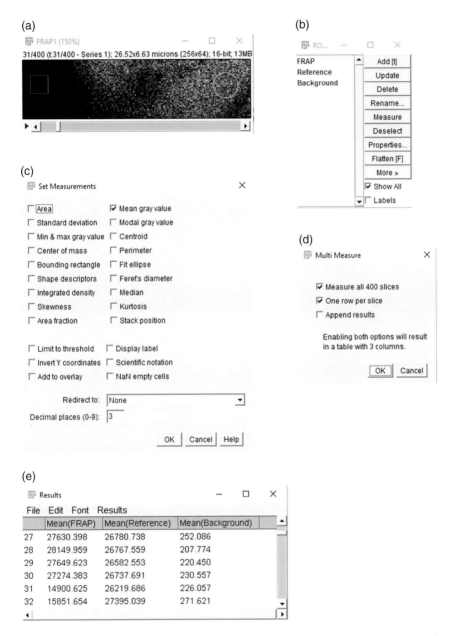

Figure 5.4 Quantification of fluorescence intensity traces in time-lapse images with ImageJ/Fiji

(a) Time-lapse image is opened in Fiji as an image stack and quantification ROIs are selected.

(b) *ROI Manager* window with three selected ROIs allows to perform quantification in multiple regions simultaneously (*More > MultiMeasure*).

(c) *Set Measurements* window.

(d) *Multi-Measure* options window.

(e) *Results* window.

- Fit normalised recoveries:
  - In the *Datasets* panel, select one dataset to fit the corresponding curve or select the name of the experiment to fit all curves that belong to it.
  - Start the fitting procedure (*Analysis > Fitting panel*).
  - In the first opened dialog (*Step 1*) select the *Fitting model*.
  - The next dialog (*Step 2*) allows modification of initial guesses for model parameters (used when previous fit was not successful). Uncheck one or several checkboxes to fix corresponding parameters during fitting. Press *OK* to run fitting.
  - Select *Parameters values* in *Show in table* drop-down list (*Settings* panel) to see fitted parameters.
  - Select *Normalised and fitted curves* in *Show on plot* drop-down list (*Settings* panel) to see fitted recoveries overlaid with experimental data points.
- Export fitted parameters as text files for further processing (*Data > Save all parameters*).

5.5 NOTES

1. Using FRAP intensity value at the first postbleach time point for normalisation and calculation of kinetics is error-prone.

   Knowledge of the bleaching depth is needed to scale the dynamic range of the FRAP recovery between 0 and 1 (Equation 5.3) and to estimate recovery halftime without fitting. Because true bleaching depth is most often not known, it is common practice to use the intensity value at the first postbleach time point as a measure of the fraction of non-bleached molecules. However, depending on the experimental settings, this estimate could be significantly biased by (a) partial recovery of fluorescence due to quick kinetics, (b) slow switching of the system from bleaching to data acquisition mode or (c) reduced sensitivity of some CLSMs' detectors for a short time after the bleaching event.

2. Use of analytical and stochastic models for analysing FRAP data:

   Differential equations and stochastic simulations are two modelling techniques applicable to FRAP data. Models based on differential equations are often represented by analytical equations that describe the FRAP signal over time in respect to specified model parameters. In these models, quantities of bleached and unbleached particles are represented by their relative concentrations. Results of such calculations are smooth recovery curves or time-dependent fluorescence distributions (for spatiotemporal recovery analysis).

In contrast, stochastic algorithms are able to mimic movement of single molecules between and within bleached and unbleached zones. For spatially resolved models, the coordinates of individual molecules are stored at each simulation time step. With these modelling procedures, synthetic FRAP recoveries are calculated by counting the number of unbleached molecules in the bleached region. Results of several simulation rounds with the same parameter set would not be fully identical and would reflect noise caused by the limited number of simulated particles.

Stochastic simulations are often used to model systems which are difficult to describe with analytical equations, e.g. diffusion limited by irregular (but known) borders, turnover of macro-molecular complexes and cytoskeleton filaments, reactions with spatially dependent rates. These simulations have much higher computational costs rendering them less attractive for repeated simulations during data fitting and larger datasets.

3. Techniques measuring the diffusion coefficient:
   Diffusion rate for proteins or other fluorescently labelled molecules can be measured by either FRAP or fluorescence correlation spectroscopy (FCS). Comparison of diffusion coefficient estimates obtained with the two methods helps to identify artefacts produced by those methods [12]. Independent diffusion measurements help to analyse recovery curves governed by both binding and diffusion. To estimate the diffusion rate for such analysis, FRAP or FCS has to be configured in one of the following ways:

   - If binding occurs only in target cell locations (e.g. organelles), diffusion measurements can be performed in other cell parts away from these places.
   - When available, a binding-incapable protein variant can be tested by FRAP or FCS to extract the diffusion rate.

5.6 CONCLUDING REMARKS

While protocols for simple qualitative comparison of recovery rates after photopleaching are relatively well established, methods for quantitative FRAP analysis are still being developed. Newly developed analysis techniques are often aimed at more accurate estimations of biophysical parameters by accounting for the experimental conditions: effective bleaching geometry, restricted cell volume, etc. Many of the recent studies suggest that such corrections are both necessary and sufficient to improve the quality of parameter estimates from single FRAP datasets. While this approach was efficient for accurate measurements of diffusion rates in spatially isotropic cell regions, they would hardly improve quantitative interpretation of other FRAP experiments when organelles with unknown or irregular shapes are bleached.

FRAP is not an exclusive technique for studying kinetics of fluorescently labelled molecules. Fluorescence correlation spectroscopy (FCS) and single particle tracking (SPT) are alternative and complementary methods for measuring molecular mobility. In contrast to FRAP, which provides only characteristic kinetic parameters for the molecular crowd in the bleached ROI, FCS and SPT are essentially single particle techniques. These techniques allow the measurement of individual molecules / complexes or discrimination between different subpopulations of molecules with different kinetic properties. Although measuring dynamics of single molecules renders these methods more precise at the molecular level, they are not well suited to studying the behaviour of a molecular crowd and are limited to fast mobility over shorter distances. Therefore photoperturbation techniques are often still the methods of choice for measuring intracellular molecular dynamics either independently or in combination with other methods. The advantage of FRAP lies in its applicability for a broad range of scientific problems and relative experimental simplicity. FCS is mainly suited to studying relatively rapid processes such as the diffusion of small molecules on the timescale up to several seconds, which is much faster then the characteristic times of some biochemical reactions. In contrast, FRAP settings can be tuned to study the dynamics of molecules ranging from subseconds to tens of minutes. While both FCS and SPT require optimisation of fluorophore concentrations in the sample, photoperturbation methods work with a larger range of fluorophore concentrations.

5.7 REFERENCES

1. Beaudouin, J. et al., *Dissecting the contribution of diffusion and interactions to the mobility of nuclear proteins.* Biophys J, 2006. **90**(6): pp. 1878–94.
2. Axelrod, D. et al., *Mobility measurement by analysis of fluorescence photobleaching recovery kinetics.* Biophys J, 1976. **16**(9): pp. 1055–69.
3. Lippincott-Schwartz, J., E. Snapp and A. Kenworthy, *Studying protein dynamics in living cells.* Nat Rev Mol Cell Biol, 2001. **2**(6): pp. 444–56.
4. Stasevich, T.J. et al., *Dissecting the binding mechanism of the linker histone in live cells: an integrated FRAP analysis.* EMBO J, 2010. **29**(7): pp. 1225–34.

5. Lai, F.P. et al., *Arp2/3 complex interactions and actin network turnover in lamellipodia.* EMBO J, 2008. **27**(7): pp. 982–92.

6. Hotulainen, P. and P. Lappalainen, *Stress fibers are generated by two distinct actin assembly mechanisms in motile cells.* J Cell Biol, 2006. **173**(3): pp. 383–94.

7. Lele, T.P. et al., *Investigating complexity of protein-protein interactions in focal adhesions.* Biochem Biophys Res Commun, 2008. **369**(3): pp. 929–34.

8. Hadzic, E. et al., *Delineating the Tes interaction site in zyxin and studying cellular effects of its disruption.* PLoS One, 2015. **10**(10): p. e0140511.

9. Liu, W. et al., *ArfGAP1 dynamics and its role in COPI coat assembly on golgi membranes of living cells.* J Cell Biol, 2005. **168**(7): pp. 1053–63.

10. Verissimo, F. et al., *A microtubule-independent role of p150glued in secretory cargo concentration at endoplasmic reticulum exit sites.* J Cell Sci, 2015. **128**(22): pp. 4160–70.

11. Weiss, M., *Challenges and artifacts in quantitative photobleaching experiments.* Traffic, 2004. **5**(9): pp. 662–71.

12. Stasevich, T.J. et al., *Cross-validating FRAP and FCS to quantify the impact of photobleaching on in vivo binding estimates.* Biophys J, 2010. **99**(9): pp. 3093–101.

13. Patterson, G.H., *Photoactivation and imaging of optical highlighter fluorescent proteins.* Curr Protoc Cytom, 2011. Chapter 12: Unit 12–23.

14. Gurskaya, N.G. et al., *Engineering of a monomeric green-to-red photoactivatable fluorescent protein induced by blue light.* Nat Biotechnol, 2006. **24**(4): pp. 461–5.

15. Rabut, G., V. Doye and J. Ellenberg, *Mapping the dynamic organization of the nuclear pore complex inside single living cells.* Nat Cell Biol, 2004. **6**(11): pp. 1114–21.

16. Koster, M., T. Frahm and H. Hauser, *Nucleocytoplasmic shuttling revealed by FRAP and FLIP technologies.* Curr Opin Biotechnol, 2005. **16**(1): pp. 28–34.

17. Lele, T.P. and D.E. Ingber, *A mathematical model to determine molecular kinetic rate constants under non-steady state conditions using fluorescence recovery after photobleaching (FRAP).* Biophys Chem, 2006. **120**(1): pp. 32–5.

18. Dunn, G.A. et al., *Fluorescence localization after photobleaching (FLAP): a new method for studying protein dynamics in living cells.* J Microsc, 2002. **205**(Pt 1): pp. 109–12.

19. Dunn, G.A. et al., *Fluorescence localization after photobleaching (FLAP).* Curr Protoc Cell Biol, 2004. Chapter 21: Unit 21–2.

20. Forster, R. et al., *Secretory cargo regulates the turnover of COPII subunits at single ER exit sites.* Curr Biol, 2006. **16**(2): pp. 173–9.

21. Costantini, L.M. et al., *A palette of fluorescent proteins optimized for diverse cellular environments.* Nat Commun, 2015. **6**: p. 7670.

22. Bancaud, A. et al., *Fluorescence perturbation techniques to study mobility and molecular dynamics of proteins in live cells: FRAP, photoactivation, photoconversion and FLIP.* Cold Spring Harb Protoc, 2010. **2010**(12): pdb top90.

23. Geverts, B., M.E. van Royen and A.B. Houtsmuller, *Analysis of biomolecular dynamics by FRAP and computer simulation.* Methods Mol Biol, 2015. **1251**: pp. 109–33.

24. Phair, R.D., S.A. Gorski and T. Misteli, *Measurement of dynamic protein binding to chromatin in vivo, using photobleaching microscopy.* Methods Enzymol, 2004. **375**: pp. 393–414.

25. Miura, K., *FrapCalc: Github release*. 2010.
26. Soumpasis, D.M., *Theoretical analysis of fluorescence photobleaching recovery experiments*. Biophys J, 1983. **41**(1): pp. 95–7.
27. Kang, M. et al., *Simplified equation to extract diffusion coefficients from confocal FRAP data*. Traffic, 2012. **13**(12): pp. 1589–600.
28. Sprague, B.L., R.L. Pego, D.A. Stavreva and J.G. McNally. 2004. Analysis of binding reactions by fluorescence recovery after photobleaching. Biophys J. **86**:3473–3495.
29. Mueller, F. et al., *FRAP and kinetic modeling in the analysis of nuclear protein dynamics: what do we really know?* Curr Opin Cell Biol, 2010. **22**(3): pp. 403–11.
30. Conrad, C., A. Wunsche, T.H. Tan, J. Bulkescher, F. Sieckmann, F. Verissimo, A. Edelstein, T. Walter, U. Liebel, R. Pepperkok and J. Ellenberg. 2011. *Micropilot: automation of fluorescence microscopy-based imaging for systems biology*. Nat Methods. **8**:246–249.
31. Tischer, C., V. Hilsenstein, K. Hanson and R. Pepperkok. 2014. Adaptive fluorescence microscopy by online feedback image analysis. Methods Cell Biol. **123**:489–503.
32. Rapsomaniki, M.A., P. Kotsantis, I.E. Symeonidou, N.N. Giakoumakis, S. Taraviras and Z. Lygerou. 2012. *easyFRAP: an interactive, easy-to-use tool for qualitative and quantitative analysis of FRAP data*. Bioinformatics. **28**:1800–1801.
33. Ulrich, M., C. Kappel, J. Beaudouin, S. Hezel, J. Ulrich, and R. Eils. 2006. *Tropical–parameter estimation and simulation of reaction-diffusion models based on spatio-temporal microscopy images*. Bioinformatics. **22**:2709–2710.

5A Case Study: Analysing COPII Turnover During ER Exit

The discussion above makes it clear that design of a particular FRAP experiment and the following data analysis steps largely depend on the properties of the investigated sample and addressed scientific question. In this case study we focus on one research example to illustrate that FRAP can provide essential kinetic information about the turnover of macro-molecular complexes.

In this study we investigate how the assembly/disassembly cycle of a COPII coat mediates protein export from ER in mammalian cells. The fully formed coat consists of identical multiprotein complexes that have multiple functions during formation of ER-derived vesicles loaded with the dedicated cargo. To perform these functions, assembly and disassembly of COPII complexes has to occur in the proper order and has to be temporarily synchronised with the formation and departure of cargo vesicles (Sato and Nakano, 2007). Although core coat components and interactions between them on ER membranes have been characterised, kinetic regulation of this process in living cells remains obscure. Several proteins have been shown to regulate secretory trafficking by interacting with COPII proteins and affecting their membrane turnover. Because COPII machinery is an essential part of protein secretion that undergoes rapid rearrangements upon cellular stimuli, more cellular components are expected to mediate COPII kinetics. To address this question systematically we experimentally estimate the influence of putative ER-exit regulators on COPII coat kinetics by FRAP. Combination of these measurements with mathematical modelling allows identification of target COPII turnover steps affected by molecular perturbations of these regulators.

5A.1 Quantitative FRAP Analysis of ER-Exit Sites

We aimed to design and conduct FRAP experiments which would provide us with a quantitative insight into the turnover of COPII complexes at ER membranes. In mammalian cells COPII assembly and export of cargo-loaded vesicles does not occur throughout the ER membrane, but on its special domains called ER-exit sites (ERES) (Figure 5.5a). Unperturbed cells typically contain several hundred ERESs that exist over a prolonged time between cell divisions. FRAP measurements showed that COPII proteins rapidly exchange between cytoplasmic and ER-bound

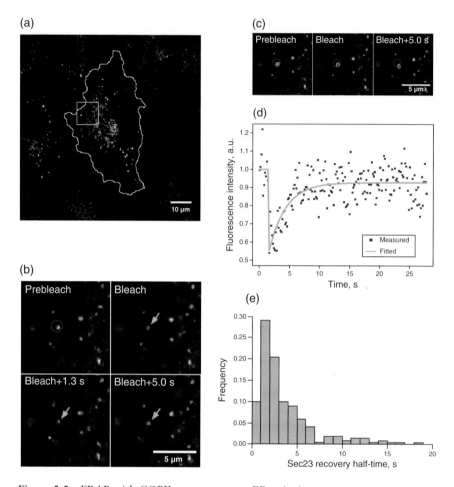

Figure 5.5 FRAP with COPII components at ER-exit sites
(a) Cell from a HeLa Kyoto derived cell line stably expressing Sec23-YFP component of the COPII coat. Bright punctae are ER-exit sites (ERES).
(b) FRAP acquisition at ERES in the region highlighted by the rectangle in (a). One
prebleach and several postbleach images from the same acquisition are shown. Overall,
10 prebleach and 170 postbleach images are acquired for each dataset with a time
resolution of 0.18 s. Five bleaching iterations are used to achieve ~50% bleaching
efficiency. The red circle is the bleached region.
(c) Quantification of FRAP recoveries at ERESs. As ERESs undergo small random
movements, their positions are tracked for the accurate recovery quantifications
(green outlines). The cytoplasmic signal is subtracted from the integrated fluorescence of the segmented punctae (Equation 5.10).
(d) Despite COPII proteins participating in multiple biochemical interactions during
turnover of the COPII coat, corresponding FRAP curves are well fitted with a single
exponential recovery curve.
(e) Distribution of estimated recovery halftimes of Sec23-YFP at ERESs. Acquisition
and analysis of such high-throughput FRAP data is feasible with automated feedback
microscopy methodology.

pools, suggesting multiple binding/unbinding cycles of COPII components at individual ERESs. However, estimates of characteristic turnover times of these proteins exhibited significant variations in the previous studies, that were likely governed by differences in experimental settings. In particular, bleaching fluorescence in bigger ROIs covering multiple ERESs resulted in significantly slower estimates of COPII turnover rates than with smaller bleaching ROIs covering individual ERESs (Forster et al., 2006; Stephens et al., 2000).

Alterations of the FRAP recovery rates upon changes of the size of the bleached spot would likely represent diffusion components in FRAP recoveries. Moreover, when several ERESs are located within the bleached region, it becomes more likely that a given bleached COPII protein will be repeatedly bound to ERESs within the bleached ROI (as in the effective diffusion regime). Therefore for unbiased measurements of the COPII assembly/disassembly cycle we currently perform all the FRAP experiments with small bleaching ROIs covering single ERESs (Figure 5.5b).

As recovery halftime for COPII proteins is in the range of one to a few seconds, square regions covering only a small part of the cell are selected for time-lapse imaging to ensure sufficiently fast frame rate (0.1–0.2 s) for quantifying fast recoveries. Imaging for 15–30 s is sufficient for most of recovery curves to reach their plateau levels.

During this period, ERESs exhibit oscillatory movements in both the vertical and horizontal directions, leading to additional complications for quantification and processing of FRAP recoveries. Slight opening of the CLSM pinhole (2.2 Airy units corresponding to 1.6 μm optical slice thickness in our experiments) makes the ERES fluorescence signal less sensitive to vertical shifts. Further pinhole opening is unfavourable due to blurring of ERES spots and decrease of their brightness relative to the surrounding cytoplasm. To compensate for horizontal movement of bleached ERESs, they are tracked automatically during data analysis.

The power of the laser line used for imaging is attenuated to 100% for bleaching. About 50% of ERES associated fluorescence is typically bleached with five bleaching iterations that take less then 0.1 s in total, allowing bleaching to be considered as an instantaneous event. The remaining ERES fluorescence enables tracking of bleached spots on the postbleach images (Figure 5.5c). Moreover, test experiments with higher bleaching depth resulted in slower recovery curves most likely because of cross-linking of coat components during bleaching.

For each image in the time course the total spot intensity is measured as an estimate of unbleached protein quantity at the analysed ERES.

The shape of the ROI is kept for quantifying ERES intensities through the time-course, to exclude variations of the quantified signal due to image noise, leading to occasional exclusion of border pixels from the segmentation results. Subtraction of the cytoplasmic contribution to the total fluorescence of segmented ERES region is performed in several steps:

1. Total intensity (I_{1tot}) and area (A_{1tot}) in the segmented ERES region are measured.
2. This region is dilated to include neighbouring cytoplasm parts except pixels belonging to neighbouring ERESs. Total intensity (I_{2tot}) and area (A_{2tot}) in the dilated region are measured.
3. The resulting ERES-associated fluorescence is calculated as:

$$I_{ERES}\left(t\right) = I_{1tot} - \frac{I_{2tot} - I_{1tot}}{A_2 - A_1} \cdot A_1 \qquad (5.10)$$

Reference curves for normalisation are measured with non-bleached ERESs on the same time-lapse images identical to quantifying FRAP curves. To avoid noise amplification when normalising with Equation 5.2, representative reference curves are generated by averaging at least 10 measured intensity traces for non-bleached ERESs. Then normalised curves are calculated using Equation 5.2 with $I_{back}(t) = 0$ because background subtraction is indirectly included in Equation 5.10.

Most of the normalised curves are well fitted with single exponential recoveries (Equation 5.4; Figure 5.5d). Data quality control includes discarding datasets with recovery fraction below 10% and with plateau levels significantly above 1. In multiple experiments, estimated recovery halftimes and immobile fractions were sensitive to molecular perturbations affecting COPII turnover (Forster et al., 2006; Verissimo et al., 2015).

5A.2 Mechanistic Insight into COPII Coat Kinetics with FRAP

Successful fitting of experimental FRAP curves with single exponential recoveries suggests that simple binding/unbinding kinetics can easily approximate membrane turnover of COPII components. In reality, assembly and disassembly of COPII complexes occurs sequentially. Inner coat components (Sec23-Sec24 heterodimer) can bind only after arrival of Sar1 protein and its activation via GDP-to-GTP exchange. The outer coat (Sec13-Sec31 heterodimers) is recruited only when the inner coat is in place. Remarkably, the probability of complex degradation increases upon its formation because both inner and outer coats stimulate GTP hydrolysis on Sar1, leading to sequential COPII disassembly.

To study the kinetics of this complex biochemical network, dedicated FRAP analysis models should integrate the turnover cycle of COPII complexes with the required level of details. Rates of FRAP recoveries predicted by such models depend on all the biochemical rate constants in the model. Separate fitting of measured FRAP curves for individual COPII components results in a plethora of parameter combinations with values distributed over several orders of magnitude. Thorough protocols for simultaneous analysis of several experimental outputs have been developed to limit the number of accepted parameter sets (Forster et al., 2006; Verissimo et al., 2015):

1. FRAP recoveries for several COPII components are predicted from the same mathematical model. The set of model parameters is considered to be appropriate only when all predicted recoveries simultaneously match the experimental data.
2. Steady-state ERES-bound fractions of imaged COPII components are predicted from the model for each parameter set. Fractions of ERES-associated fluorescence are measured from collected cell images and compared to steady-state model predictions to filter out improper parameter sets.
3. Because of the interdependencies of binding and unbinding rates for different coat components, the membrane turnover of each of them depends on the abundance of COPII proteins in the cell. Biochemical measurements of relative expression levels for these proteins help to choose between parameter sets which satisfy both steady-state and kinetic FRAP data (Forster et al., 2006).
4. For rather complex models the number of parameter combinations could pass through multiple experimental restrictions. These sets can be used to build distributions of parameter values that are narrower for better defined parameters. Quantitative comparison of these distributions for wild-type cells and cells in which COPII kinetics was perturbed, highlights the steps in COPII turnover affected by these perturbations (Verissimo et al., 2015).

The described combination of quantitative FRAP, steady-state fluorescence analysis and mathematical modelling has previously provided novel insight into the regulation of COPII-mediated ER exit in the cellular context. Interactions of cargo molecules with COPII proteins were shown to support vesicle formation by stabilising the COPII coat (Forster et al., 2006). Perturbation of lipid composition of ER membranes resulted in reduced cargo export kinetics accompanied by alterations of COPII assembly and disassembly rates (Runz et al., 2006). Further on, the p150[glued] protein, which is part of the dynactin complex, was shown to contribute to the coat stability and to support cargo load into formed carriers (Verissimo et al., 2015). Further development of models based on FRAP and supporting microscopy measurements should provide understanding of the functions of other regulatory molecules mediating secretory flux at ERESs.

5A.3 Automated FRAP at ERESs

Systematic FRAP-based analysis of COPII kinetics at ERESs requires a sufficient number of recoveries to be collected and analysed. Although acquisition of a single recovery takes around 30 s, the experimental throughput is drastically decreased by the need to manually switch between low-zoom prescan and FRAP acquisition settings and by the time spent in finding appropriate cells and ERESs for bleaching. Furthermore, a lot of collected datasets are excluded from the analysis when significant ERES movements lead to poor quantifications of recovery curves.

To overcome this limitation, FRAP protocols for ERESs have been combined with adaptive feedback microscopy (Tischer et al., 2014). Overnight experiments (~15 h) with our experimental settings result in 500–600 collected FRAP datasets. About 35% collected recoveries pass quality control criteria, which are essentially the same as for manual FRAP, but incorporated into automated analysis procedures.

Averaging of recovery halftimes for several hundred automatically acquired and processed FRAP datasets resulted in values comparable with results of manual experiments. This sample size also permitted more comprehensive analysis of distributions of recovery rates at individual ERESs. Interestingly, individual $t_{1/2}$ values were broadly and asymmetrically distributed for several COPII components (Figure 5.5e). While most recoveries were relatively fast, a percentage of FRAP acquisitions (typically 5–10%) resulted in much slower recoveries leading to increased $t_{1/2}$ values.

Further automated experiments need to be implemented to determine whether the differences between COPII turnover rates at individual ERESs have functional meaning. Previously observed correlations between coat stability and efficiency of carrier formation allow the hypothesis that fast and slow coat turnover correspond to different vesicle production rates for corresponding EREss. As mathematical modelling and quantitative analysis were previously performed with average halftime values, further update of modelling procedures is required to interpret observed differences in COPII turnover in terms of biochemical rates of coat (dis-) assembly.

5A.4 References

Forster, R., M. Weiss, T. Zimmermann, E.G. Reynaud, F. Verissimo, D.J. Stephens and R. Pepperkok. 2006. *Secretory cargo regulates the turnover of COPII subunits at single ER exit sites.* Curr Biol. **16**:173–179.

Runz, H., K. Miura, M. Weiss and R. Pepperkok. 2006. *Sterols regulate ER-export dynamics of secretory cargo protein ts-O45-G.* EMBO J. **25**:2953–2965.

Sato, K. and A. Nakano. 2007. *Mechanisms of COPII vesicle formation and protein sorting.* FEBS Lett. **581**:2076–2082.

Stephens, D.J., N. Lin-Marq, A. Pagano, R. Pepperkok and J.P. Paccaud. 2000. *COPI-coated ER-to-Golgi transport complexes segregate from COPII in close proximity to ER exit sites.* J Cell Sci. **113** (Pt 12):2177–2185.

Tischer, C., V. Hilsenstein, K. Hanson and R. Pepperkok. 2014. *Adaptive fluorescence microscopy by online feedback image analysis.* Methods Cell Biol. **123**:489–503.

Verissimo, F., A. Halavatyi, R. Pepperkok and M. Weiss. 2015. *A microtubule-independent role of p150glued in secretory cargo concentration at endoplasmic reticulum exit sites.* J Cell Sci. **128**:4160–4170.

6

Co-Localisation and Correlation in Fluorescence Microscopy Data

Dylan Owen, George Ashdown, Juliette Griffié and Michael Shannon
Department of Physics and Randall Division of Cell and Molecular Biophysics, King's College London, UK

6.1 INTRODUCTION

The development of fluorescence microscopy has unravelled the complexity and importance of molecular spatiotemporal organisation in cells [1, 2]. Today, the interacting partners in many signalling pathways have been well identified. What has become obvious is that many molecules (e.g. proteins) appear in multiple signalling pathways, often within the same cell. The question therefore arises as to how cells regulate signalling and avoid issues of non-specificity and crosstalk between pathways [3]. It is now broadly recognised that cellular machinery relies essentially on the relocation of key signalling proteins. Put another way, it is the spatiotemporal regulation of lipids and proteins that allows multiple signalling pathways to cooperate within a single cell. Cell proliferation, migration and activation for instance, have extensively been shown to depend on such reorganisation in time and space at the molecular level [4–9]. In these contexts, objectively quantifying the molecular co-localisation,

Standard and Super-Resolution Bioimaging Data Analysis: A Primer, First Edition.
Edited by Ann Wheeler and Ricardo Henriques.
© 2018 John Wiley & Sons Ltd. Published 2018 by John Wiley & Sons Ltd.

complex formation, diffusion behaviour and active transport is key to improving our understanding of many cellular processes [10].

Fluorescence microscopy is a powerful tool for identifying the sub-cellular localisation of signalling molecules. Even in its most basic implementation, it provides submicron spatial resolution [11]. Images can be acquired quickly – even at video rate. In addition, the technique is highly specific to the protein of interest when extrinsic fluorescent tags are used and the method is also minimally invasive – allowing live cell, or even in vivo imaging. Depending on the implementation, 3D resolution is possible – so called 'optical sectioning' [12], multicolour microscopy is well established and new super-resolution microscopy techniques have pushed the spatial resolution into the nanometre range [13–17].

To address the requirements of investigating molecular co-localisation, complex formation, diffusion and flow, new analysis tools have been developed in parallel with the emergence of fluorescence microscopy. This chapter will begin with an exploration of co-localisation analysis for conventional fluorescence microscopy images. These are images consisting of arrays of pixels each one having a digital value proportional to the fluorescence intensity at that location. While caveats (e.g. variations in quantum yield or illumination profile) exist, this fluorescence intensity is generally taken to be proportional to the fluorophore concentration. These types of images are typical of those acquired from wide-field epifluorescence microscopes, confocal [12] or multiphoton microscopes [18] or total internal reflection fluorescence (TIRF) systems [19, 20]. Each has advantages and disadvantages in terms of speed, optical section-ing, signal-to-noise and so on, but all are ultimately limited in resolution by the diffraction limit of light to around 200 nm.

See section on co-localisation in super resolution localisation micros-copy and Chapter 8 please. Super-resolution is a group of techniques which circumvent the diffraction limit of light and allow spatial resolution into the nanometre range. There are several different implementations but the three most common are stimulated emission depletion micros-copy (STED) [17, 21–23], structured illumination microscopy (SIM) [16, 24, 25] and single molecule localisation microscopy (SMLM) [15, 26–28]. While STED and SIM produce images that are the same format as conventional microscopes (i.e. arrays of pixels), SMLM does not, instead producing pointillist data sets. SMLM imaging has recently exploded in popularity for the simple reason that it combines nanometre resolution while maintaining the specificity of fluorescent labelling coveted by biologists. Maps of single molecule positions are useful for studying the clustering of molecules – a spatiotemporal phenomenon crucial for cell signalling [29–32]. While biochemistry and diffraction limited microscopy have allowed us to understand the importance of clustering and spatial regulation in cells, they only provide averaged population information on a phenomenon that changes quickly at tiny length scales. Super-resolution localisation microscopy gives us access to information on this nanometre length scale, thus providing a wealth of information previously hidden behind the diffraction limit.

Moving on from localisation and co-localisation, the final part of this chapter will discuss the other aspect of spatiotemporal organisation – molecular dynamics. Imaging and quantifying molecular transport at subcellular levels provides key insights into mechanistic and dynamic interactions that occur under physiological conditions. The principal mechanism by which molecules move around cells is diffusion, typically described in the context of Brownian motion [33–37]. However, the cell interior is a complex environment which presents obstacles, semi-permeable barriers, molecular crowding and aspects of active transport and flow. When faced with such an environment, molecules deviate from Brownian motion towards more anomalous behaviour. Here we outline the principles, theories and applications of two widely established fluorescence techniques for measuring and quantifying such molecular behaviour. Fluorescence correlation spectroscopy (FCS) [23, 38–40] allows researchers to detect sparsely distributed molecules and to quantify molecular concentrations and diffusion coefficients. Image correlation spectroscopy (ICS) [41] provides a tool for quantifying the speed and direction of molecular populations during imaging of larger fields of view – that is, spatially resolved measurements of molecular transport [42, 43].

6.2 CO-LOCALISATION FOR CONVENTIONAL MICROSCOPY IMAGES

Co-localisation software is now widely available to study images obtained from two-colour fluorescence microscopy [44]. Dual-colour imaging is achieved by simultaneously targeting two molecules of interest with fluorophores of different wavelengths, thus allowing the study of their interactions in the same cell. Conventional fluorescence microscopy is defined by a limit in resolution of around 200 nm, resulting from the wave nature of light. Although new microscopy techniques such as super-resolution fluorescence microscopy (see section on co-localisation in super resolution localisation microscopy and Chapter 8), have managed to circumvent this resolution limit, in this current section we will focus on routinely used diffraction limited microscopes such as total internal reflection fluorescence (TIRF) or confocal setups. These imaging techniques rely on optical sectioning or filtering of out-of-focus emitted light. Here, we will describe state-of-the-art quantification strategies for co-localisation in data acquired by these microscopy techniques.

Because of the diffraction limit associated with the use of conventional microscopy, the distribution of the fluorophores is seen as a blurry pixelated image. Indeed, the principal effect of the limit in resolution is that the output image is an imperfect representation of the fluorophore distribution in which the intensity is proportional to the density of fluorophores. It does not provide an exact location for each fluorophore within the sample, hence exact co-localisation cannot be extracted from such data sets. More precisely, each fluorophore appears as an intensity function spread over multiple pixels, commonly called the point spread function (PSF). Hence, fluorophores in close proximity will appear as overlapping PSFs, forcing us to redefine co-localisation in the context of resolution limited images. Thus, the co-localisation of two molecules is seen as a local similarity in the intensity profile of both channels, to be interpreted as an indication of close proximity and the possibility of molecular interactions. If it is necessary to know precisely whether two fluorophore are in extreme close proximity ('bound'), techniques such as FRET may be used [45–47].

As a more general and reproducible illustration, we will focus on the case of two proteins of interest, protein A (PA) tagged with fluorophore A (FA) and protein B (PB) with fluorophore B (FB), imaged with a conventional fluorescence microscope such as a confocal system. The resulting output is two images, in two different channels (one channel for each fluorophore). The first (trivial) way to visualise co-localisation is to overlay the images resulting from both channels. Supposing no chromatic aberration or channel crosstalk, and using an appropriate virtual colour scale (for instance, red for PA-FA, green for PB-FB and yellow for pixels with both), it is possible to determine visually if there is co-localisation in specific areas of the image. This strategy nonetheless presents various disadvantages. Although it allows a fast and easy visualisation of co-localisation, it does not provide any robust quantification to study the specificity of PA and PB interactions from one condition to the other. This lack of comparable quantification limits the use of this method in the context of complex biological studies. Also, it is highly dependent on the signal-to-noise ratio, and requires the intensity profile to be similar in both channels. In fact, in the case of two different profiles of intensity and protein distributions, the overlay of the images could lead to an incomplete or even incorrect interpretation of the co-localisation between channels 1 and 2. It is therefore usually used only for visualisation purposes and systematically associated with quantification analysis methods presented in following subsections.

Two main strategies have been developed to answer the requirements of quantification in the case of dual colour conventional microscopy images. The first strategy relies on intensity correlation coefficients and the second is based on object recognition. In the context of intensity correlation coefficient based analysis, a broadly used representation of co-localisation between channel 1 and channel 2 is the intensity scatter plot. By representing the intensity in channel 1 ($I1$ on the x axis) as a function of the intensity in channel 2 ($I2$ on the y axis) for each equivalent pixel of the input images, the scatter plot provides valuable information not only on the co-localisation status but also on the kind of interaction existing. Indeed, in the case of two completely co-localising proteins of interest (when targeting the same protein with two different fluorophores as a positive control for instance), the intensity profiles over the images are very similar. Hence high and low intensity pixels coincide in each channel. Thus, the resulting scatter plot displays a linear distribution which can easily be fitted, the slope depending on the ratio between $I1$ and $I2$. The resulting fit will coincide with the points of the scatter plot (Figure 6.1a). If there is partial co-localisation, the slope of the scatter plot will also be positive but with a wider cloud of points (Figure 6.1b). In the case of exclusion, one protein causes the absence of the other and the trend of the points will be negative (Figure 6.1c). The extreme case of this is complete exclusion where $I1 > 0$ will imply $I2 = 0$. Similarly, $I2 > 0$ will imply $I1 = 0$ (Figure 6.1d). Although these cases provide very recognisable scatter plots, when dealing with real data sets, the image is most likely to be predominated by partial co-localisation in the presence of noise. In these cases, Pearson's coefficient is routinely used as a reliable descriptor of co-localisation. This correlation coefficient consists of estimating the quality of the linear fit extracted from the scatter plot. More precisely, it will estimate the correlation between the linear approximation and the points on the scatter plot. The Pearson's coefficient, R_p, is defined as

$$R_p = \frac{\sum_n \left(I1 - \overline{I1}\right) \times \left(I2 - \overline{I2}\right)}{\sqrt{\sum_n \left(I1 - \overline{I1}\right)^2 \times \sum_n \left(I2 - \overline{I2}\right)^2}}$$

with $I1$ and $I2$ representing the intensities in channels 1 and 2. R_p varies from -1 for exclusion to $+1$ for complete co-localisation.

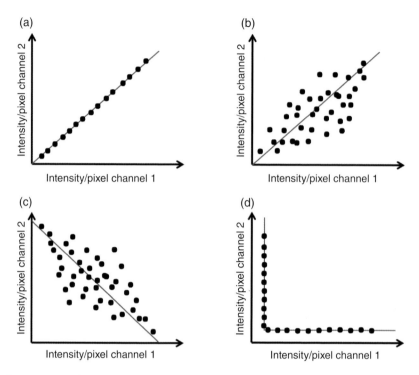

Figure 6.1 Schematic intensity scatter plots between two fluorescence channel intensities for the case of (a) perfect co-localisation, (b) partial co-localisation, (c) partial exclusion and (d) total exclusion.

Another correlation-based approach that relies on Pearson's coefficient calculation, proposed by Van Steensel, tests for proximity and repetitive patterns rather than direct co-localisation. The idea behind this technique is to determine the Pearson's coefficient as a function of δx and δy; calculating the shift in the x and y spatial dimensions [48]. This strategy is particularly powerful in the case of large structures where PA and PB do not strictly co-localise, as describe in Figure 6.2, or in the presence of chromatic aberration. In this technique, the image is shifted along x and y directions, and Pearson's coefficient calculated for each shift. If indeed repetitive patterning exists between PA and PB, they will be highlighted as peaks in the plot of R_p as a function of δx for instance.

Although used extensively, Pearson's coefficient presents some strong disadvantages. It is indeed a good indicator of the presence of co-localisation, in particular, complete or strong co-localisation, but in the case of partial co-localisation the results can be ambiguous. Manders et al. has shown that Pearson's coefficient will be similar in a vast range of scenarios, leading to the possible misinterpretation of data sets [49]. For instance, they calculate

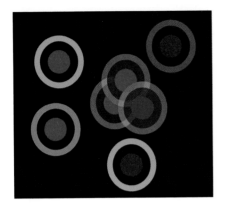
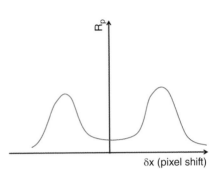

Figure 6.2 Schematic of Van Steensel's method. If two fluorophores (red and green in this example) do not show complete co-localisation but do show a relation, calculation of Pearson's correlation coefficient, R, in the presence of shifts in the position of one image relative to the other (in this case along the x axis) will display peaks, with the position of the peaks indicating the spatial separation of the fluorophores.

the Pearson's coefficient values in two very distinct scenarios in the absence of noise. In the first case, both channels have similar signal intensity and distribution and 25% of PA co-localises with 33% of PB. In the second example, the signal intensity and distribution are drastically different between the channels and only 8% of PA co-localises with 75% of PB. It appears that, for both cases, the Pearson's coefficient method provides very similar results (0.22 vs 0.23). Taken together, these disparities suggest that to obtain a more precise and comparable co-localisation coefficient, it should be calculated relative to the intensity in one of the channels.

In 1993, Manders et al. proposed an alternative to Pearson's method which, by providing a relative coefficient for each channel, is designed to palliate the limitations of the Pearson's coefficient and is still considered to be the best method for the study of complex interactions [49]. M1 and M2 correspond to the ratio of the number of co-localising pixels for each channel when pixels below a specified intensity level have been set to zero and are defined as

$$M1 = \frac{\sum_n I1 * I2}{\sum_n I1^2}$$ the Manders' coefficient relative to channel 1.

$$M2 = \frac{\sum_n I1 * I2}{\sum_n I2^2}$$ the Manders' coefficient relative to channel 2.

M1 and M2 values vary from 0 (for the exclusion scenario) to 1 (for the complete co-localisation scenario). Because the two channels are dealt with independently, the Manders' coefficients remain relevant in the case of very different contribution in co-localisation from channel 1 and 2. In particular, for the two scenarios described previously, the resulting Manders' coefficients are significantly different and illustrate the co-localisation status with precision (M1 = 25 and M2 = 33 for the first case, M1 = 8 and M2 = 75 for the second). It has to be noted that these results have been obtained on simulated data sets in the absence of noise. Indeed, the obvious limitation of this strategy is its sensitivity to noise and hence the necessity to set a channel specific threshold for M1 and M2. The choice of a threshold can introduce errors in interpretation because what should be considered as background can vary from one condition to the other in a set of biological samples. Indeed, although new strategies have been developed to try to determine an adequate threshold [50], there is a remaining risk of filtering out valuable information at low intensities. As both coefficients, Pearson's and Manders', are easy to extract from the image, we recommend that they be used in combination, to allow the correct interpretation of the data sets.

The main limitations of any intensity correlation coefficient based analysis are the lack of visualisation and local quantification. Indeed, because these techniques rely on the intensity scatter plot, they can be viewed as providing an estimation of the co-localisation between channel 1 and 2, pixel by pixel over the defined region of interest, and therefore lack structural information. By contrast, object based co-localisation analysis deals with each channel independently to identify objects, before overlaying the two resulting images to search for overlapping objects. Many strategies are available for object recognition and image segmentation, depending on the kind of structures studied. A simplistic example of the basics behind object recognition would be to add an intensity threshold to each channel. Costes et al.'s thresholding method can be used in this context too. Every pixel above that threshold can therefore be considered as being part of an object. To identify and delimit one object, the basic idea would be to use a connectivity condition for the attribution of pixels to an object (each pixel above threshold connected together by an edge, or an edge or a corner, will be attributed to the same object). The edge of the object can be determined from pixels in the object which have a side connected to a non-object pixel. The last step consists of overlaying both binary output maps and searching for object overlap which would suggest the interaction of PA and PB. To quantify the overlap, we need to take into account the diffraction limit associated with the detection setup. This is because, in conventional diffraction-limited microscopy, real-world objects appear in the image as a convolution of that object with the microscope point-spread-function.

6.2.1 Co-Localisation in Super-Resolution Localisation Microscopy

In the SMLM variants of super-resolution microscopy, multiple proteins of interest can be labelled with fluorophores conjugated to specific antibodies or using different fluorescent proteins, then microscopy techniques like stochastic optical reconstruction microscopy (STORM) [27, 28] and photoactivatable light microscopy (PALM) [15, 26] can be used to achieve spatial separation of emitters by exploiting photoswitching. In multicolour mode, frames can be cycled from one laser line to the next or the channels imaged simultaneously using a dichroic mirror. Raw data consists of thousands of frames each with a subset of PSFs. The centres of these PSFs are then localised and expressed as x-y coordinates, typically with a precision of 10–30 nm using one of several available algorithms [51–54]. The localised centroids are then summed to form a final map of the positions of the emitters. Unlike conventional microscopy images, data generated from SMLM techniques are in the form of a list of x and y (and z) coordinates, each with an associated localisation precision value. The quantitation of such spatial point patterns (SPPs) has led to the development and application of 'cluster analysis' tools, such as Ripley's K-function [55, 56] and pair correlation [57, 58]. While both techniques have been used widely, they may require user input and calibration data that strongly influences the outcome and does not take into account the localisation precision. Thus, recently, a model-based Bayesian approach to cluster analysis was developed to analyse molecular clustering in single colour SPPs [59]. Co-clustering of biological complexes is known to be crucial for regulation. For example, the association of active Src family kinases has been investigated by confocal microscopy and biochemistry, and is known to be a functionally important regulator downstream of the T cell receptor and integrin adhesome [5, 30, 60]. With multicolour imaging modalities now available in direct STORM [61] and PALM [62], combined with the concurrent development of the analysis techniques for multicolour SPPs [63, 64], this kind of information is now investigable at the nanoscale.

Many different imaging modalities exist, but most rely on the use of spectrally distinct fluorophores activated by a particular set of laser wavelengths or the use of a dichroic mirror and subsequent spectral unmixing.

One problem with these methods, interchangeably called crosstalk or bleed-through, refers to the improper channel assignment of the probe due to overlap of their excitation or emission spectra. Any interchange of species between two channels will increase the subsequent level of correlation/co-localisation. To remedy this, suitable filters are employed to avoid the majority of spectral overlap. Crosstalk can be reduced to < 1% in post-processing by applying a photon filter so that only those fluorophores activated by the correct laser line and therefore emitting a high amount of signal are registered. Correction of bleed-through can be achieved using stochastic methods such as alteration of laser power [65], or quantitatively during image analysis [66]. Corrections for crosstalk routinely achieve a level that makes it undetectable in subsequent analyses [67].

A second issue that arises is chromatic aberration. This results in two identically located molecules appearing separated. Achromatic lenses are used to prevent chromatic aberration, but are less effective at nanometre length scales. One way to correct for errors on this length scale is to image multicolour fluorescent beads [63, 67] or DNA rulers bound to fluorophores [68], such that PSFs emanating from emitters on the same bead in multiple channels are known distance apart. Chromatic errors can then be corrected for by using these reference positions plotted in a matrix, which can reduce error to a negligible level.

Finally, linkage error describes a limit on resolution based on the size of the fluorescently conjugated probe and therefore the distance of the emitter from the protein target – these problems are compounded when looking at two targets whose spatial co-localisation might be important at the nanoscale. Commonly used antibodies have a size of ~15 nm. As a consequence in a primary plus secondary antibody labelling schema, there will be up to a 30 nm gap between emitter and target. Several solutions have been touted for this problem: in STORM these include the use of nanobodies [69], SNAP-tags [70] and aptamers [71].

While the imaging considerations are similar to traditional fluorescence microscopy (with a few caveats), localisation microscopy derived pointillist data requires a completely different approach to dealing with what it means to be co-localised; inherently, there is no longer any 'intensity' information, so efforts to quantitate co-localisation have to be based on the proximity of points. The simplest method for comparing two sets of coordinate data is to employ the nearest neighbour technique [72, 73], used frequently in ecology, geography and astronomy. Simply put, nearest neighbour gives each point a value based on its proximity to each other point. To extend this to co-localisation analysis, each localisation of species A is given a nearest neighbour value based on how close it is to a localisation of species B [74]. In practical biological terms, a working value for perfect co-localisation is defined by imaging molecule A tagged with two suitable fluorophores. The problem with this method is that it is a coarse quantitation, and gives only a general answer relating to how co-clustered the points in a given region appear to be.

The pair-correlation technique (PC) [75] calculates an autocorrelation function for the point pattern at different length scales, which measures the probability density around each point. The pair correlation function quantifies the probability of finding another point from a reference point relative to a random distribution. When proteins are in clusters, the correlation value will be greater than 1.

In the cross-correlation (two-colour data) case, the measured cross-correlation, $c(r)^{peaks}$, data is simply fit to the following equation, where $c(r)^{PSF}$ is the cross-correlation of the PSFs of the two fluorophores:

$$c(r)^{peaks} = \left(A^{-r/\xi} + 1\right) * c(r)^{PSF}$$

An alternative method 'coordinate based co-localisation (CBC) was conceived by Malkusch et al. [63], which relies on counting the number of localisations of species B around each localisation A, $N_{Ai,B}(r)$, within a given radius, r, and the same is done for self-clustering by $N_{Ai,A}(r)$. This is then normalised to both the number of localisations within the largest observed distance (R_{max}) and area for each species.

$$D_{A,A}(r) = \frac{N_{A,A}(r)}{N_{A,A}(R_{max})} \frac{R_{max}^2}{r^2}$$

$$D_{A,B}(r) = \frac{N_{A,B}(r)}{N_{A,B}(R_{max})} \frac{R_{max}^2}{r^2}$$

The two distributions are compared and the Spearman rank correlation coefficient is calculated to give a value for co-clustering of A with B. Thus, each localisation is attributed an individual co-localisation value calculated by the Spearman's rank coefficient between the linearised distribution functions of each population (A and B) around said localisation. The authors then applied a weighting based on the distance of a given localisation from its nearest neighbour, in order to prevent false positive co-localisation values being generated when the nearest neighbour is very far away. Thus, co-localisation is measured on a scale of −1 for anti-correlated to 0 for randomly distributed and 1 for correlated for each species separately.

Rossy et al. provide an alternative technique, which relies on a Getis and Franklin version of the Ripley's K technique [64, 76] (Figure 6.3). In this method, the number of molecules within a specified radius of each molecule is counted. This number is then normalised to the total molecular density of the analysed region to provide a measure of how 'clustered' that molecule is – the L value. When applied to a co-localisation setting, the level of cross-clustering value, L(r)cross is introduced, where points from population B are counted around each point A, and vice versa (thus this technique is an asymmetric method). L(r)cross can then be plotted against L(r) and the percentage of clusters co-localised with clusters as well as the free population of monomers can be extracted for each channel by applying thresholds to L(r) and L(r)cross. The L(r) and L(r)cross value is given by

$$L(r)_j = \sqrt{A\sum_{i=1}^{n}\left(\frac{\delta_{ij}}{n}\right)/\pi} \text{ where } \delta_{ij} = \left\{\begin{array}{l} 1 \text{ is } d_{ij} < 1 \\ else\, 0 \end{array}\right\}$$

where A is the analysed area, n is the total number of points in the area and d_{ij} is the distance between points i and j. The length scale employed, r, can be altered depending on the size-scale of objects in the data, for example a small radius for small clusters, and a large radius for large clusters. Like CBC, this type of analysis is asymmetric: it could delineate for example a situation where all clusters of A are all associated with clusters of B, whereas some clusters of B are free of A [76].

Super-resolution microscopy is a young field – it is only now that the imaging techniques are becoming widespread and applicable within a biological setting. Biophysicists are now able to use these techniques to reliably acquire two sets of colour data to investigate multiple protein targets. As with the single colour scenario, sophisticated quantitation methods are only just catching up with the imaging. One of the future directions for these techniques is to extend multicolour quantitation to the third dimension [77–80], which will afford greater depth to questions relating to membrane proximal protein trafficking as a spatial regulator of signalling pathways. These approaches are now also being applied to live-cell quantification [81], but unlike the quantification of structures in which SMLM excels, quantification of dynamics is the forte of the correlation approaches – fluorescence correlation and image correlation spectroscopy.

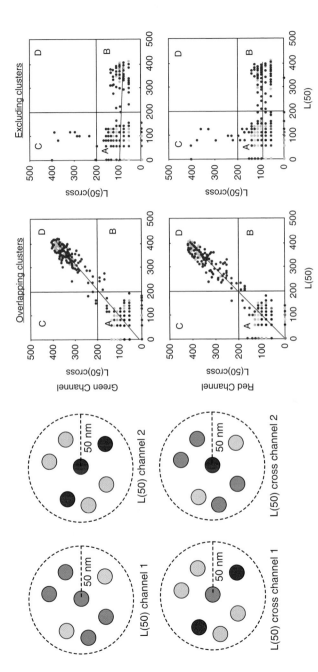

Figure 6.3 (a) Method for co-cluster analysis using Getis' variant of Ripley's K-function. Red molecules are counted within a radius centred on each green molecule, and vice versa, to give the L(r)cross values (in this case the radius = 50). The L(r) values are then plotted against L(r)cross for each channel. The resulting plots can then be divided into quadrants and, by applying thresholds, the strength of the correlation can be determined to assess co-clustering. Colours indicate how many molecules in a region of interest had that specific combination of L(r) and L(r)cross value.

6.2.2 Fluorescence Correlation Spectroscopy

Fluorescence correlation spectroscopy (FCS) [82, 83] monitors spontaneous fluctuations in fluorescence intensity within a static detection volume (Figure 6.4). By using a confocal or two-photon microscope this detection volume is around 1 fL, and intensity fluctuations mirror fluorophores moving in and out of this volume.

Through analysis of the amplitude and duration of these fluctuations as they pass through the observation volume, information can be extracted on single molecule diffusion rates, reaction kinetics and molecular concentrations down to the pico-molar range. This gives researchers the ability to collect information at the single molecule scale with a relatively standard microscope system. FCS can also span temporal ranges from microseconds to minutes. The measurement volume (Figure 6.4) is described mathematically by the PSF of the microscope. Characterisation of the PSF (or beam radius, e^{-2}) is important as its shape varies from system to system and this variation will have an effect on the generated fluctuation profiles. One way of calibrating the measurement volume is to generate a sample of known concentration and which exhibits a known

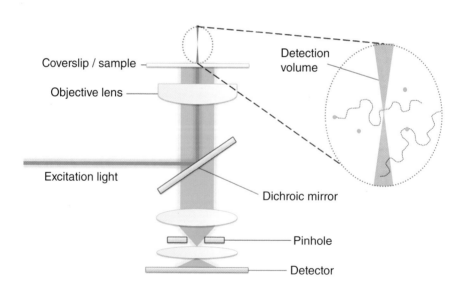

Figure 6.4 Basic setup and principle of fluorescence correlation spectroscopy. Excitation light is directed through the high NA objective into the sample, creating the excitation volume. Fluorescent molecules travelling into this volume are excited and their fluorescence is collected back through the objective onto a photon detector.

diffusion coefficient due to the simple Brownian motion of its molecules. The PSF and therefore observation volume of the system is estimated by

$$PSF(r,z) = I_0 e^{-2r^2/\omega_{xy}^2} e^{-2z^2/\omega_z^2}$$

Where PSF(r, z) is the radial and axial position, I_0 is the intensity peak, ω_{xy} and ω_z are radial and axial radii where $\omega_z > \omega_{xy}$. For confocal microscopes employing a pinhole size of 1 Airy unit, the PSF ω_{xy} is limited by the diffraction limit to around 200 nm, with ω_z up to six times greater.

Once a data set has been recorded, the data is analysed via an autocorrelation. An intensity reading I at timepoint t is compared with another intensity reading at timelag (τ) from t. After applying the timelag, the output autocorrelation defines the similarities between the observations at time t and time $t+\tau$ (Figure 6.5). The autocorrelation is described as

$$G(\tau) = \frac{\left(\delta I[t]\delta I[t+\tau]\right)}{\left(I[t]\right)^2} = \frac{\left(I[t]I[t+\tau]\right)}{\left(I[t]\right)^2} - 1$$

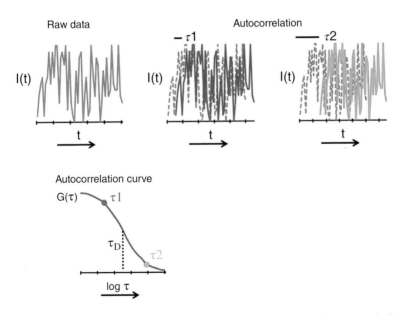

Figure 6.5 By correlating a data set with itself over varying time lags (τ), underlying characteristics can be extracted. Intensities measured at shorter time lags $(\tau1)$ tend to show similarities giving a higher correlation. Over longer time lags (autocorrelation $\tau2$) the data is less similar, resulting in a reduced correlation. Once the correlation function $(G(\tau))$ is plotted over time, the point at half the maximum amplitude gives the correlation time τ_D.

Where $G(\tau)$ is the correlation function, $\delta I\,[t]$ is the deviation in intensity from the mean intensity. The denominator here is the most commonly used as it is normalised, meaning the correlation at $\tau = 0$, $G(0)$, is inversely related to the number of particles within the excitation volume. The correlated data is then usually fitted by a nonlinear least squares algorithm; here the Levenberg–Marquardt algorithm is shown:

$$x_v^2 = \frac{\sum_i \left(\left[y(x_i) - y_i \right]^2 / \delta_i^2 \right)}{v = p}$$

Where $y(x_i)$ are the fitting values, y_i are the data point values and σ_i is the standard deviation of point i. And v is the number of significant data points, while p the number of free parameters.

After quantification of the raw data, the shape of the curve distribution can be used to determine the observed molecular dynamics. For standard Brownian motion of molecules in 3D, subjected to thermal diffusion, the autocorrelation curve is theoretically given by

$$G(\tau) = G(0) \frac{1}{\left(1 + \left[\dfrac{\tau}{\tau_D} \right] \right)\left(1 + \beta^{-2} \left[\dfrac{\tau}{\tau_D} \right] \right)^{\frac{1}{2}}} + G(\infty)$$

where β is the axial to radial ratio of the observation volume ω_z/ω_{xy}, and the correlation time, τ_D is the average dwell time of the molecule within the observation volume. $G(0)$ is the initial autocorrelation amplitude and $G(\infty)$ is the correlation amplitude at very long lags. τ_D can be used to recover the diffusion coefficient using

$$D = \frac{\omega_{xy}^2}{4\tau_D}$$

In cases where diffusion is anomalous (e.g. restricted, directed or trapped) by forces other than thermal, α (the anomalous diffusion coefficient) becomes a free function:

$$G(\tau) = G(0) \frac{1}{\left(1 + \left[\dfrac{\tau}{\tau_D} \right]^{\alpha} \right)\left(1 + \beta^{-2} \left[\dfrac{\tau}{\tau_D} \right]^{\alpha} \right)^{\frac{1}{2}}} + G(\infty)$$

Both these conditions assume that the photophysics of the probe are stable. Corrections for fluorescent signal fluctuations (i.e. dark or triplet state transitions) can also be calculated by fitting more complex functions. These variables allow changes in molecular behaviour to be observed, including molecular interactions: where slower diffusions exhibit a rightward shift in the autocorrelation curve due to increased correlation times. Higher concentrations (a larger average number of particles within the focal volume) are detected as a reduction in height of the plotted autocorrelation curve at $G(0)$.

Many biological questions have been investigated using FCS, from the original paper by Elson and Magde (1974) [92] demonstrating the chemical dynamics of DNA-drug intercalation, through fundamental applications such as the effects of fluorophore concentration on the molecules mobility in solution, to investigations of molecule interactions and membrane dynamics [40, 84, 85]. Here, we focus on the cellular applications. When studying diffusion in near-2D cellular environments such as the plasma membrane, it is important to consider the behavioural diversity exhibited by lateral diffusion. The previous section outlined equations that covered the diffusion and observation of molecules in a 3D model, but fluorophores in the plasma membrane are confined to 2D motion, so

$$G(\tau) = G(0)\left(1 + \tau/\tau_D\right)^{-1} = G(0)\left(1 + \frac{4D_\tau}{s^2}\right)^{-1}$$

where s is the beam waist (replacing the 3D volume) and the diffusion coefficient is provided by $\tau_D = s^2/4D$. When using FCS to study the more confined dynamics within the plasma membrane, certain advantages of the technique can become limitations. As diffusion times are $\times 10^4$ slower within the membrane, longer acquisition times are required to collect enough data for reliable analysis. Another consideration when setting up an FCS experiment is that it relies on knowing the size of the observation volume. Distortions to the volume can be caused by refractive index mismatch between the coverslip and sample and, if not accounted for, will lead to inaccuracies when analysing the results. It must also be assumed that the z position of the membrane is at the beam waist (its narrowest radius) which is difficult to know. Technical extensions which do not rely on a priori knowledge of the excitation volume (i.e. calibration free) include Z-scan FCS [86], and scanning FCS.

When discussing molecular dynamics within a cellular environment, one should consider that the majority of these are governed by interactions with other molecules; as such it is also important to take these into account when investigating biophysics. Fluorescence cross-correlation spectroscopy (FCCS) [87–90] permits the study of two different fluorescent populations, recorded on separate detectors. By separating two fluorescent populations that exhibit different emission wavelengths through the use of optical filters, correlation time (τ_D) data can be analysed separately by autocorrelation for diffusion coefficients, but also, through cross-correlation any molecular interactions that are occurring can be analysed. FCCS has extracted information on a range of molecular interactions and binding rates, including protein co-localisation (such as during signalling complex formation), mobility and enzymatic activity [91, 92].

The setup of an FCCS experiment, especially for live-cell imaging, can be difficult to achieve because the two excitation volumes must be completely co-registered and the emission spectra of the two fluorophores should be well separated to allow for minimal crosstalk between the two detectors. FCCS has the advantage of eliminating after-pulsing artefacts which can be present when using certain types of detector. In a similar vein, time-resolved FCS or fluorescence lifetime correlation spectroscopy uses a pulsed laser source to measure lifetimes of single molecules, allowing separation of species based on their lifetimes rather than their spectra [93]. FLCS relies on recording two timescales; the photon arrival time relative to the excitation pulse (picosecond range) and the photon arrival relative to the start of the experiment (microsecond range).

Until recently, the size of the excitation volume (and hence the spatial and temporal resolution of the measurements) had been limited to the diffraction limit of the microscope system. By employing a stimulated emission (STED) [17] beam to FCS measurements, Eggeling et al. reduced the excitation volume of the system thereby increasing the spatial resolution in x, y [23, 94]. As correlation functions require the detection of a small number of molecules to extract intensity fluctuations smaller volumes lead to more sensitive measurements revealing, for example, millisecond scale molecular trapping. So STED-FCS has the additional benefit of improving the temporal resolution by detecting dynamic, single molecules with greater resolution. To increase the data gleaned from single emitters to include spatial and temporal information, scanning FCS was created, which does not park a spot on a single area of the sample but scans in a raster fashion or spins the observation point around the sample [95–97]. The additional information from what is essentially tens or hundreds of individual FCS observation points gives researchers not only a greater understanding of the sample through greater n-numbers, but also spatial and temporal information.

Several limitations arise when using scanning or spinning FCS, including cell motion, bleaching of immobile fractions, and large fluctuations in the average intensity from varying signal due to sample labelling heterogeneities. Also, the scan/spin cycle must be fast enough so that measured particles at point x do not completely leave the area in the following 'frame' to allow tracking of the signal decay. Scanning FCS data can be shown as a carpet or kymograph, with x, y representing the points of a scan and time respectively. FCS is a highly specialised method for investigating the concentration, mobility and interactions of fluorescent molecules within a biological specimen or solution. The tools required and the acquisition of data is relatively straightforward, but the analysis and interpretation of results still requires specialist skills and understanding.

6.2.3 *Image Correlation Spectroscopy*

Image correlation spectroscopy (ICS) [41, 98] encompasses a series of techniques based on the detection of fluorescence fluctuations that occur in space, through time. These analysis methods correlate the changes in signal intensity based on the directed or diffusive flow of fluorescent molecules in or through a region of interest. ICS was originally an extension of scanning-FCS, widening the correlative power through x, y space. Using confocal or wide-field imaging to sample the cell of interest, this has advantages over single-point correlation techniques when the diffusion is slow or the molecular populations sparse. ICS uses the intensity of each pixel as a concentration measurement of the fluorophores, allowing quantification of flow velocities within the sample and can be used to image whole-cell flow dynamics. Unlike FCS, which has sampling speeds in the picosecond range, ICS is limited to the imaging speed of the microscope, which may miss faster diffusing populations, but does allow the tracking of slow and even stochastic movements of whole populations through time.

Fundamentally, the ICS technique measures pixel intensity fluctuations through space, but this can be extended to investigate fluctuations through time [43, 99], where pixel pairs that have matching spatial and temporal lags are grouped into a correlation function. When the data set is analysed, this shows spatial correlation as a function of time. This output depends on the molecular dynamics and fluorophore photophysics. When the time lag (τ) is zero, each image is autocorrelated with itself and so the maximum intensity distribution which is displayed as a

Gaussian distribution (the approximate PSF profile of the microscope system) is at 0 (centre of plot), representing no flow.

For isotropic diffusion (i.e. non-directed flow) the correlation function $r_{ab}(\xi,\eta,\tau)$ decays symmetrically in all directions through time, while dynamic samples with longer time lags ($\tau > 0$) evolve different correlation functions ($\tau 0$ to $\tau 2 \neq \tau 0$ to $\tau 3$). For detected flow, the correlation peak shifts and translates from the zero spatial lags (central) origin as a function of time lags. The distance from zero of the correlation function is found by locating the maximum $r_{ab}(\xi,\eta,\tau)$ peak fitted to a Gaussian function with the fit parameters where g_{ab} is the time dependent amplitude:

$$r'_{ab}\left(\Delta x, \Delta y, \Delta t\right) = g_{ab}\left(\xi - \Delta x, n - \Delta y, \Delta t\right)\exp\left(-\frac{\left(\xi - \Delta x\right)^2 + \left(n - \Delta y\right)^2}{\omega\left(\Delta t\right)^2}\right) + r_\infty$$

where the Gaussian function is fitted to the spatial Δx and Δy and temporal amplitude peak g_{ab}, the time dependent peak is e^{-2} radius ω and r_o is the spatial offset variable.

Any immobile objects within the detection ROI remain centred within the correlation function, with their amplitude proportional to their density. During slow flow conditions, or when a sample is not imaged over a long enough temporal range, the mobile fraction may not leave this immobile peak and the mobile population can be obscured as this also originates at the centre. There are, however, several ways of filtering out this immobile signal including Fourier methods.

For data acquisition, sampling at the Nyquist–Shannon criteria is recommended; that is, a pixel size should be at most half the diameter of the PSF of the microscope in size. It is also important to generate enough signal above the noise of the system to allow the ICS algorithm to detect molecules, however this can result in excessive photobleaching. As the entire sample, or at least a large subset of the sample, is often imaged to capture slow and quasi-static populations it is important to minimise the photobleaching effects, through minimising laser powers. As ICS methods rely on fluctuations in fluorescence they are limited to studying spatially heterogeneous samples, or at least heterogeneously labelled samples. This can become an issue as many labelling methods for fluorescence imaging use protein overexpression to study molecular dynamics; this should be taken into account when preparing samples for ICS analysis. Another limitation of spatiotemporal image correlation spectroscopy (STICS) is the assumption that the sample is near-2D; as such, acquisitions should be taken in as thin an optical plane as possible, such as by employing TIRF microscopy. As STICS relies on correlating fluorescent distributions within a finite ROI (subregion), individual frame acquisition times should be kept as short as possible for higher temporal resolutions. Greater temporal resolutions, along with short dwell times between frames, will ensure fluorescent fluctuations (movement of molecules) do not shift too quickly and move out of this ROI between frames, which would mean flow information is lost due to under-sampling.

One of the principal advantages of STICS is to establish whether fluorescence is exhibiting a net flow out of the observation areas, i.e. to extract directionality. The spatial information extracted from the autocorrelation equation is therefore combined with the temporal correlation:

$$r^I{}_{ab}\left(\xi,\eta,\Delta t\right) = \frac{1}{N-\Delta t} \sum_{t=1}^{N-\Delta t}$$

In order to maximise the spatial resolution, STICS can be combined with wide-field super-resolution techniques such as SIM giving pixel sizes of the reconstructed images down to 30 nm [100, 101].

Much like FCCS, two-colour correlation can also be employed with ICS techniques (ICCS [102] and STICS [43, 103]). The equation below extends the correlative function to include two channels a and b:

$$r_{ab}\left(\xi,\eta,\tau\right) = \frac{\left(\left[\delta i_a\left(x,y,t\right)\delta i_b\left(x+\xi,y+\eta,t+\tau\right)\right]_{xy}\right)_t}{\left(i_a\right)_t\left(i_b\right)_{t+\tau}}$$

For two-channel imaging it is important to align the microscope correctly prior to imaging, and align the data after acquisition, before correlation. Depending on the software being used, this can be done automatically or manually through ImageJ plugins, and some of these require fiducials within the imaged region to achieve accurate alignment. When imaging in series rather than simultaneously, consideration should be given to the length of time required to collect the two images, as some systems require on the order of seconds to switch between the desired image settings, by which time the imaged populations may not be seen to correlate with each other.

For higher resolution correlation spectroscopy, particle ICS (PICS) [104] and time-resolved STICS (trSTICS) [105] give researchers the ability to analyse SPT and SMLMS data respectively and can achieve nanometre and millisecond spatial and temporal resolutions respectively. PICS has the advantage over classic SPT of not breaking down if multiple trajectories overlap.

One disadvantage of all ICS techniques to this point, is the assumption that the detected fluorescence fluctuations come from transport through the ROI, and not from intrinsic emission variations; however, the photophysics of fluorophores is known to fluctuate due to environmental factors and photobleaching. k-space ICS (kICS) [106] offers several advantages over real-space ICS methods by analysing data after transformation to Fourier space. It is insensitive to blinking or bleaching fluorophores and the PSF of the system does not need to be measured before analysis. ICS and its extensions provide a valuable insight into the molecular dynamics and interactions of cellular environments. Acquiring the data requires little expansion beyond that provided by commercially available microscope systems, and results are relatively intuitive compared with other quantitative techniques.

6.3 CONCLUSION

Fluorescence microscopy has been a boon to the life sciences in recent years. One of the major challenges is extracting quantitative information on biological systems from the images generated from increasingly sophisticated hardware and labelling strategies. One of the most common requirements by far is to quantify the co-localisation of molecules from multicolour microscopy images. In conventional fluorescence microscopy, there are many well established techniques – principal among them the Pearson's correlation coefficient and the Manders co-localisation coefficients with a number of practical guides available for their implementation [44]. This field is well established and the methods are unlikely to evolve greatly in future. In the emerging field of super-resolution microscopy, in which the data is pointillist rather than pixelated, new tools for quantifying co-localisation are only now being developed. The pointillist nature of the data, being described mathematically as a spatial point pattern, has forced the rethinking of what it means for two molecules to be co-localised in such data sets. For example, since molecules are almost never distributed randomly within a real biological sample, the methods of quantifying co-localisation directly overlap with tools designed to quantify molecular clustering. In such studies, it must be made clear whether the goal is to quantify the co-localisation of molecules or clusters as these results have different biological implications.

As new statistical tools emerge, we will gain new insight into how to accurately describe molecular interactions in cells. It is likely that the co-localisation analysis will directly mirror developments in the methodology for structure identification in pointillist data sets.

While co-localisation analysis and structural image analysis go hand-in-hand, other methods excel at extracting dynamic information from image time-series. A family of techniques of this type is correlation analysis – which itself has also been applied to the co-localisation problem. Correlation analyses extract how quickly a signal changes over space, time or both. While these methods are themselves well-established for quantifying biophysical mechanisms such as diffusion, trapping and flow (active transport), they are now evolving to deal with the data sets from the latest developments in microscope hardware [23, 81, 100, 105, 107]. Many of the advancements are being made to deal with the pointillist nature of SMLM data, multicolour analysis and 3D sets. As these developments continue, they will allow the quantification of molecular dynamics on smaller and smaller length scales and shorter and shorter timescales, revealing previously hidden biophysical processes.

6.4 ACKNOWLEDGMENTS

Dylan Owen acknowledges funding from the European Research Council (FP7 starter grant 337187) and Marie Curie Career Integration grant 334303. Michael Shannon was supported by the King's Bioscience Institute and the Guy's and St Thomas' Charity Prize PhD Programme in Biomedical and Translational Science.

6.5 REFERENCES

1. Michalet, X. et al., *The power and prospects of fluorescence microscopies and spectroscopies.* Annual Review of Biophysics and Biomolecular Structure, 2003. **32**(1): p. 161–182.
2. Yuste, R., *Fluorescence microscopy today.* Nature Methods, 2005. **2**(12): p. 902–904.
3. Lewitzky, M., P.C. Simister and S.M. Feller, *Beyond 'furballs' and 'dumpling soups' – towards a molecular architecture of signaling complexes and networks.* FEBS Letters, 2012. **586**(17): p. 2740–2750.
4. Kanchanawong, P. et al., *Nanoscale architecture of integrin-based cell adhesions.* Nature, 2010. **468**(7323): p. 580–584.
5. Williamson, D.J. et al., *Pre-existing clusters of the adaptor Lat do not participate in early T cell signaling events.* Nature Immunology, 2011. **12**(7): p. 655–662.

6. Karunarathne, W.K.A. et al., *Optical control demonstrates switch-like PIP3 dynamics underlying the initiation of immune cell migration.* Proceedings of the National Academy of Sciences, 2013. **110**(17): p. E1575–E1583.
7. Cebecauer, M. et al., *Signalling complexes and clusters: functional advantages and methodological hurdles.* Journal of Cell Science, 2010. **123**(3): p. 309–320.
8. Lidke, D.S. and B.S. Wilson, *Caught in the act: quantifying protein behaviour in living cells.* Trends in Cell Biology, 2009. **19**(11): p. 566–574.
9. Vögler, O. et al., *Membrane interactions of G proteins and other related proteins.* Biochimica et Biophysica Acta, 2008. **1778**(7–8): p. 1640–1652.
10. Magee, T. et al., *Lipid rafts: cell surface platforms for T cell signaling.* Biological Research, 2002. **35**(2): p. 127–131.
11. Lichtman, J.W. and J.-A. Conchello, *Fluorescence microscopy.* Nature Methods, 2005. **2**(12): p. 910–919.
12. Conchello, J.-A. and J.W. Lichtman, *Optical sectioning microscopy.* Nature Methods, 2005. **2**(12): p. 920–931.
13. Schermelleh, L., R. Heintzmann and H. Leonhardt, *A guide to super-resolution fluorescence microscopy.* Journal of Cell Biology, 2010. **190**(2): p. 165–175.
14. Cella Zanacchi, F. et al., *Live-cell 3D super-resolution imaging in thick biological samples.* Nature Methods, 2011. **8**(12): p. 1047–1049.
15. Betzig, E. et al., *Imaging Intracellular Fluorescent Proteins at Nanometer Resolution.* Science, 2006. **313**: p. 1642–1645.
16. Gustafsson, M.G.L., *Surpassing the lateral resolution limit by a factor of two using structured illumination microscopy.* Journal of Microscopy, 2000. **198**(2): p. 82–87.
17. Hell, S.W. and J. Wichmann, *Breaking the diffraction resolution limit by stimulated emission: Stimulated-emission-depletion fluorescence microscopy.* Optics Letters, 1994. **19**(11): p. 780–782.
18. Centonze, V.E. and J.G. White, *Multiphoton excitation provides optical sections from deeper within scattering specimens than confocal imaging.* Biophysical Journal, 1998. **75**(4): p. 2015–2024.
19. Axelrod, D., *Cell surface contacts illuminated by total internal refection fluorescence.* Journal of Cell Biology, 1981. **89**: p. 141–145.
20. Axelrod, D., *Total internal reflection fluorescence microscopy in cell biology.* Traffic, 2001. **2**(11): p. 764–774.
21. Harke, B. et al., *Resolution scaling in STED microscopy.* Optics Express, 2008. **16**(6): p. 4154–4162.
22. Hein, B., K.I. Willig and S.W. Hell, *Stimulated emission depletion (STED) nanoscopy of a fluorescent protein-labeled organelle inside a living cell.* Proceedings of the National Academy of Sciences, 2008. **105**(38): p. 14271–14276.
23. Eggeling, C. et al., *Direct observation of the nanoscale dynamics of membrane lipids in a living cell.* Nature, 2009. **457**: p. 1159–1163.
24. Shao, L. et al., *Super-resolution 3D microscopy of live whole cells using structured illumination.* Nature Methods, 2011. **8**(12): p. 1044–1046.
25. York, A.G. et al., *Resolution doubling in live, multicellular organisms via multifocal structured illumination microscopy.* Nature Methods, 2012. **9**(7): p. 749–754.

26. Hess, S.T., T.P.K. Girirajan and M.D. Mason, *Ultra-high resolution imaging by fluorescence photoactivation localization microscopy.* Biophysical Journal, 2006. **91**(11): p. 4258–4272.
27. Rust, M.J., M. Bates and X. Zhuang, *Sub-diffraction-limit imaging by stochastic optical reconstruction microscopy (STORM).* Nature Methods, 2006. **3**(10): p. 793–796.
28. Heilemann, M. et al., *Subdiffraction-resolution fluorescence imaging with conventional fluorescent probes.* Angewandte Chemie International Edition, 2008. **47**(33): p. 6172–6176.
29. Hartman, N.C., J.A. Nye and J.T. Groves, *Cluster size regulates protein sorting in the immunological synapse.* Proceedings of the National Academy of Sciences, 2009. **106**(31): p. 12729–12734.
30. Rossy, J. et al., *Conformational states of the kinase Lck regulate clustering in early T cell signaling.* Nature Immunology, 2013. **14**(1): p. 82–89.
31. Kenworthy, A.K., *Nanoclusters digitize Ras signalling.* Nature Cell Biology, 2007. **9**(8): p. 875–877.
32. Owen, D.M. et al., *Sub-resolution lipid domains exist in the plasma membrane and regulate protein diffusion and distribution.* Nature Communications, 2012. **3**: p. 1256.
33. Saffman, P.G. and M. Delbruck, *Brownian motion in biological membranes.* Proceedings of the National Academy of Sciences, 1975. **72**(8): p. 3111–3113.
34. Owen, D.M. et al., *Quantitative microscopy: protein dynamics and membrane organisation.* Traffic, 2009. **10**(8): p. 962–71.
35. Simson, R. et al., *Structural mosaicism on the submicron scale in the plasma membrane.* Biophysical Journal, 1998. **74**(1): p. 297–308.
36. Heinemann, F., Sven K. Vogel and P. Schwille, *Lateral membrane diffusion modulated by a minimal actin cortex.* Biophysical Journal, 2013. **104**(7): p. 1465–1475.
37. Sankaran, J. et al., *Diffusion, Transport, and cell membrane organization investigated by imaging fluorescence cross-correlation spectroscopy.* Biophysical Journal, 2009. **97**(9): p. 2630–2639.
38. Schwille, P., J. Korlach and W.W. Webb, *Fluorescence correlation spectroscopy with single-molecule sensitivity on cell and model membranes.* Cytometry, 1999. **36**(3): p. 176–182.
39. Bacia, K. et al., *Fluorescence correlation spectroscopy relates rafts in model and native membranes.* Biophysical Journal, 2004. **87**(2): p. 1034–1043.
40. Wawrezinieck, L. et al., *Fluorescence correlation spectroscopy diffusion laws to probe the submicron cell membrane organization.* Biophysical Journal, 2005. **89**(6): p. 4029–4042.
41. Petersen, N.O. et al., *Quantitation of membrane receptor distributions by image correlation spectroscopy: concept and application.* Biophysical Journal, 1993. **65**(3): p. 1135–1146.
42. Wiseman, P.W. et al., *Spatial mapping of integrin interactions and dynamics during cell migration by Image Correlation Microscopy.* Journal of Cell Science, 2004. **117**(23): p. 5521–5534.
43. Hebert, B., S. Costantino and P.W. Wiseman, *Spatiotemporal image correlation spectroscopy (STICS) theory, verification, and application to protein velocity mapping in living CHO cells.* Biophysical Journal, 2005. **88**(5): p. 3601–3614.

44. Bolte, S. and F.P. Cordlière, *A guided tour into subcellular colocalization analysis in light microscopy*. Journal of Microscopy, 2006. **224**: p. 213–232.
45. Jares-Erijman, E.A. and T.M. Jovin, *FRET imaging*. Nature Biotechnology, 2003. **21**(11): p. 1387–1395.
46. Jares-Erijman, E.A. and T.M. Jovin, *Imaging molecular interactions in living cells by FRET microscopy*. Current Opinion in Chemical Biology, 2006. **10**(5): p. 409–416.
47. Zeug, A. et al., *Quantitative Intensity-based FRET approaches A comparative snapshot*. biophysical journal, 2012. **103**(9): p. 1821–1827.
48. van Steensel, B. et al., *Partial colocalization of glucocorticoid and mineralocorticoid receptors in discrete compartments in nuclei of rat hippocampus neurons*. Journal of Cell Science, 1996. **109**(4): p. 787–792.
49. Manders, E.M.M., F.J. Verbeek and J.A. Aten, *Measurement of colocalization of objects in dual-color confocal images*. Journal of Microscopy, 1993. **169**(3): p. 375–382.
50. Costes, S.V. et al., *Automatic and quantitative measurement of protein-protein colocalization in live cells*. Biophysical Journal, 2004. **86**(6): p. 3993–4003.
51. Small, A. and S. Stahlheber, *Fluorophore localization algorithms for super-resolution microscopy*. Nature Methods, 2014. **11**(3): p. 267–279.
52. Henriques, R. et al., *QuickPALM: 3D real-time photoactivation nanoscopy image processing in ImageJ*. Nature Methods, 2010. **7**(5): p. 339–340.
53. Ovesný, M. et al., *ThunderSTORM: a comprehensive ImageJ plug-in for PALM and STORM data analysis and super-resolution imaging*. Bioinformatics, 2014. **30**(16): p. 2389–2390.
54. Wolter, S. et al., *rapidSTORM: accurate, fast open-source software for localization microscopy*. Nature Methods, 2012. **9**(11): p. 1040–1041.
55. Ripley, B.D., *Modelling spatial patterns*. Journal of the Royal Statistical Society B, 1977. **39**: p. 172–192.
56. Owen, D.M. et al., *PALM imaging and cluster analysis of protein heterogeneity at the cell surface*. Journal of Biophotonics, 2010. **3**(7): p. 446–454.
57. Sengupta, P. et al., *Probing protein heterogeneity in the plasma membrane using PALM and pair correlation analysis*. Nature Methods, 2011. **8**(11): p. 969–975.
58. Sengupta, P. and J. Lippincott-Schwartz, *Quantitative analysis of photoactivated localization microscopy (PALM) data sets using pair-correlation analysis*. BioEssays, 2012. **34**(5): p. 396–405.
59. Rubin-Delanchy, P. et al., *Bayesian cluster identification in single-molecule localization microscopy data*. Nature Methods, 2015. **12**(11): p. 1072–1076.
60. Bakker, G.J. et al., *Lateral mobility of individual integrin nanoclusters orchestrates the onset for leukocyte adhesion*. Proceedings of the National Academy of Sciences, 2012. **109**(13): p. 4869–4874.
61. van de Linde, S. et al., *Multicolor photoswitching microscopy for subdiffraction-resolution fluorescence imaging*. Photochemical & Photobiological Sciences, 2009. **8**(4): p. 465–469.
62. Shroff, H. et al., *Dual-color superresolution imaging of genetically expressed probes within individual adhesion complexes*. Proceedings of the National Academy of Sciences, 2007. **104**(51): p. 20308–20313.

63. Malkusch, S. et al., *Coordinate-based colocalization analysis of single-molecule localization microscopy data.* Histochemistry and Cell Biology, 2012. **137**(1): p. 1–10.

64. Rossy, J. et al., *Method for co-cluster analysis in multichannel single-molecule localisation data.* Histochemistry and Cell Biology, 2014. **141**(6): p. 605–612.

65. Bates, M. et al., *Multicolor Super-Resolution Imaging with Photo-Switchable Fluorescent Probes.* Science, 2007. **317**: p. 1749–1753.

66. Kim, D. et al., *Bleed-through correction for rendering and correlation analysis in multi-colour localization microscopy.* Journal of Optics, 2013. **15**(9): p. 094011.

67. Churchman, L.S. et al., *Single molecule high-resolution colocalization of Cy3 and Cy5 attached to macromolecules measures intramolecular distances through time.* Proceedings of the National Academy of Sciences of the United States of America, 2005. **102**(5): p. 1419–1423.

68. Schmied, J.J. et al., *Fluorescence and super-resolution standards based on DNA origami.* Nature Methods, 2012. **9**(12): p. 1133–1134.

69. Ries, J. et al., *A simple, versatile method for GFP-based super-resolution microscopy via nanobodies.* Nature Methods, 2012. **9**(6): p. 582–584.

70. Klein, T. et al., *Live-cell dSTORM with SNAP-tag fusion proteins.* Nature Methods, 2011. **8**(1): p. 7–9.

71. Opazo, F. et al., *Aptamers as potential tools for super-resolution microscopy.* Nature Methods, 2012. **9**(10): p. 938–939.

72. Clark, P.J. and F.C. Evans, *Distance to Nearest neighbor as a measure of spatial relationships in populations.* Ecology, 1954. **35**(4): p. 445–453.

73. Perry, G.L.W., *SpPack: spatial point pattern analysis in Excel using Visual Basic for Applications (VBA).* Environmental Modelling & Software, 2004. **19**(6): p. 559–569.

74. Koyama-Honda, I. et al., *Fluorescence imaging for monitoring the colocalization of two single molecules in living cells.* Biophysical Journal, 2005. **88**(3): p. 2126–2136.

75. Sengupta, P., T. Jovanovic-Talisman and J. Lippincott-Schwartz, *Quantifying spatial organization in point-localization superresolution images using pair correlation analysis.* Nature Protocols, 2013. **8**(2): p. 345–354.

76. Getis, A. and J. Franklin, *Second-Order Neighborhood Analysis of Mapped Point Patterns.* Ecology, 1987. **68**(3): p. 473–477.

77. Owen, D.M. et al., *Quantitative analysis of three-dimensional fluorescence localization microscopy data.* Biophysical Journal, 2013. **105**(2): p. L05–L07.

78. Huang, B. et al., *Three-dimensional super-resolution imaging by stochastic optical reconstruction microscopy.* Science, 2008. **319**(5864): p. 810–813.

79. Juette, M.F. et al., *Three-dimensional sub-100 nm resolution fluorescence microscopy of thick samples.* Nature Methods, 2008. **5**(6): p. 527–529.

80. Xu, K., H.P. Babcock and X. Zhuang, *Dual-objective STORM reveals three-dimensional filament organization in the actin cytoskeleton.* Nature Methods, 2012. **9**(2): p. 185–188.

81. Stone, M.B. and S.L. Veatch, *Steady-state cross-correlations for live two-colour super-resolution localization data sets.* Nature Communications, 2015. **6**.

82. Elson, E.L. and D. Magde, *Fluorescence correlation spectroscopy: Conceptual basis and theory.* Biopolymers, 1974. **13**(1): p. 1–27.

83. Kim, S.A., K.G. Heinze and P. Schwille, *Fluorescence correlation spectroscopy in living cells.* Nature Methods, 2007. 4(11): p. 963–973.

84. Lenne, P.-F. et al., *Dynamic molecular confinement in the plasma membrane by microdomains and the cytoskeleton meshwork.* EMBO Journal, 2006. 25: p. 3245–3256.

85. Wenger, J. et al., *Diffusion analysis within single nanometric apertures reveals the ultrafine cell membrane organization.* Biophysical Journal, 2007. 92(3): p. 913–919.

86. Humpolíčková, J. et al., *Probing diffusion laws within cellular membranes by Z-scan fluorescence correlation spectroscopy.* Biophysical Journal, 2006. 91(3): p. L23–L25.

87. Schwille, P., F.J. Meyer-Almes and R. Rigler, *Dual-color fluorescence cross-correlation spectroscopy for multicomponent diffusional analysis in solution.* Biophysical Journal, 1997. 72(4): p. 1878–1886.

88. Bacia, K., S.A. Kim and P. Schwille, *Fluorescence cross-correlation spectroscopy in living cells.* Nature Methods, 2006. 3(2): p. 83–89.

89. Bacia, K. and P. Schwille, *A dynamic view of cellular processes by in vivo fluorescence auto- and cross-correlation spectroscopy.* Methods, 2003. 29(1): p. 74–85.

90. Bacia, K. and P. Schwille, *Practical guidelines for dual-color fluorescence cross-correlation spectroscopy.* Nature Protocols, 2007. 2(11): p. 2842–2856.

91. Slaughter, B.D., J.W. Schwartz and R. Li, *Mapping dynamic protein interactions in MAP kinase signaling using live-cell fluorescence fluctuation spectroscopy and imaging.* Proceedings of the National Academy of Sciences, 2007. 104(51): p. 20320–20325.

92. Weidemann, T. et al., *Single cell analysis of ligand binding and complex formation of interleukin-4 receptor subunits.* Biophysical Journal, 2011. 101(10): p. 2360–2369.

93. Rüttinger, S. et al., *On the resolution capabilities and limits of fluorescence lifetime correlation spectroscopy (FLCS) measurements.* Journal of Fluorescence, 2010. 20(1): p. 105–114.

94. Mueller, V. et al., *STED Nanoscopy reveals molecular details of cholesterol- and cytoskeleton-modulated lipid interactions in living cells.* Biophysical Journal, 2011. 101(7): p. 1651–1660.

95. Honigmann, A. et al., *Scanning STED-FCS reveals spatiotemporal heterogeneity of lipid interaction in the plasma membrane of living cells.* Nature Communications, 2014. 5.

96. Digman, M.A. et al., *Measuring fast dynamics in solutions and cells with a laser scanning microscope.* Biophysical Journal, 2005. 89(2): p. 1317–1327.

97. Digman, M.A. et al., *Fluctuation correlation spectroscopy with a laser-scanning microscope: exploiting the hidden time structure.* Biophysical Journal, 2005. 88(5): p. L33–36.

98. Comeau, J.W.D., S. Costantino and P.W. Wiseman, *A Guide to Accurate Fluorescence Microscopy colocalization measurements.* Biophysical Journal, 2006. 91(12): p. 4611–4622.

99. Brown, C.M. et al., *Probing the integrin-actin linkage using high-resolution protein velocity mapping.* Journal of Cell Science, 2006. 119(24): p. 5204–5214.

100. Ashdown, George W. et al., *Molecular flow quantified beyond the diffraction limit by spatiotemporal image correlation of structured illumination microscopy data*. Biophysical Journal, 2014. **107**(9): p. L21–L23.
101. Ashdown, G. et al., *Cortical actin flow in T cells quantified by spatio-temporal image correlation spectroscopy of structured illumination microscopy data*. Journal of Visualized Experiments, 2015(106): p. e53749.
102. Srivastava, M. and N.O. Petersen, *Image cross-correlation spectroscopy: A new experimental biophysical approach to measurement of slow diffusion of fluorescent molecules*. Methods in Cell Science, 1996. **18**(1): p. 47–54.
103. Toplak, T. et al., *STICCS Reveals matrix-dependent adhesion slipping and gripping in migrating cells*. Biophysical Journal, 2012. **103**(8): p. 1672–1682.
104. Semrau, S. and T. Schmidt, *Particle image correlation spectroscopy (PICS): retrieving nanometer-scale correlations from high-density single-molecule position data*. Biophysical Journal, 2007. **92**(2): p. 613–621.
105. Pandžić, E., J. Rossy and K. Gaus, *Tracking molecular dynamics without tracking: image correlation of photo-activation microscopy*. Methods and Applications in Fluorescence, 2015. **3**(1): p. 014006.
106. Kolin, D.L., D. Ronis and P.W. Wiseman, *k-Space image correlation spectroscopy: A method for accurate transport measurements independent of fluorophore photophysics*. Biophysical Journal, 2006. **91**(8): p. 3061–3075.
107. Godin, A.G. et al., *Revealing protein oligomerization and densities in situ using spatial intensity distribution analysis*. Proceedings of the National Academy of Sciences, 2011. **108**(17): p. 7010–7015.

7

Live Cell Imaging: Tracking Cell Movement

Mario De Piano[1], Gareth E. Jones[2] and Claire M. Wells[3]

[1] *Division of Cancer Studies, King's College London, UK*
[2] *Randall Division of Cell & Molecular Biophysics, King's College London, UK*
[3] *School of Cancer and Pharmaceutical Sciences, King's College London, UK*

7.1 INTRODUCTION

In the human body, the ability of cells to actively migrate is an important biological parameter in tissue development, repair and homeostasis (Cordlières et al., 2013). In addition to these essential physiological events, migration is also an important characteristic in several diseases, such as chronic inflammatory disease, vascular disease and metastasis. The acquisition of a distinct migratory potential is a hallmark of malignant transformation in epithelial cells and is a critical step in the metastatic cascade which ultimately leads to their dissemination into foreign non-native tissue. Thus, cell migration is a highly dynamic phenomenon that has become central in the study of cancer in biomedical research. Using imaging techniques such as time-lapse microscopy, we are able to observe the spatiotemporal behaviour of a migrating cell, either randomly or directionally, in response to intracellular and extracellular signalling cues (Meijering et al., 2012). Depending on the number of conditions and the overall time needed to observe the given effects of each condition, this

Standard and Super-Resolution Bioimaging Data Analysis: A Primer, First Edition.
Edited by Ann Wheeler and Ricardo Henriques.
© 2018 John Wiley & Sons Ltd. Published 2018 by John Wiley & Sons Ltd.

method can allow for the production of relatively large data sets. These data sets are not usually a continuous acquisition of events from start to finish, but images taken sequentially in time at a desired rate depending upon the dynamic processes being investigated (Miura, 2005). Once these microscopy images are acquired they may need to be pre-processed in order to be readable in certain software packages and suitable for the algorithms intended for tracking analysis.

7.2 SETTING UP A MOVIE FOR TIME-LAPSE IMAGING

The design and initial setup of the experiment is an essential prerequisite for obtaining good feedback data which will allow the observer to make accurate conclusions about the hypothesis tested. The first step of a migration assay involves seeding cells, either onto pre-treated coverslips, or more frequently, into the wells of a plastic tissue culture plate. In theory, any treated flat-bottomed plastic cell culture plate ranging from 6 to 96 wells can be used in 2D migration assays, assuming the plate sits flat on the stage. In practice when performing an experiment with several conditions and thus with several wells, the smaller the diameter of the well the more likely you are to encounter fluctuations in light intensity due to the depth of the meniscus along the rim of the well. This curvature at the air to medium interface leads to a loss in contrast as a result of the displacement of the light path and thus it may be more suitable to use wells with a diameter of 22 mm or above so that several fields can be imaged more centrally in the well (Pawley, 2010). When performing a migration assay experiment, it is best to seed the cells at a relatively low density. An environment where the cells are sparsely seeded is important as it allows the individual cells the freedom to move across the terrain without being influenced by surrounding cells. Crowding cells can alter the behaviour, directionality and speed of individual cells through contact inhibition (Middleton and Sharp, 1984).

Once the cells have been seeded and allowed to properly adhere and spread, image acquisition via time-lapse microscopy can begin. To maintain cells under normal homeostatic conditions over an extended period of time cells should be kept in a humid, warm, CO_2 gassed media environment. Time-lapse microscopes used for studying cell migration are usually modified to be fitted with a temperature thermostat and a chamber which can simulate the optimal conditions needed by the cells. The ecology of the chamber should be set up at least 2 hours before starting the experiment. Typically 37 °C is the optimal temperature for mammalian cells. Many microscopes are fitted with a CO_2 supply, but if not the culture medium of the cells can be buffered to allow for pH stability over long periods of time (Jain et al., 2012).

Once the plate has been fitted onto the microscope stage, the requirements of the assay can be set in the imaging software. The choice of objective used in this assay is partially dependent on the size of the cells and how motile they are likely to be. If the cells have been seeded relatively sparsely then a ×10 objective might be needed to obtain a field with an acceptable number of cells to track (ideally n ≥ 30 per condition). It is also wise to carefully check that your phase contrast microscope is properly aligned. If the phase plate and condenser annulus are misaligned then this could lead to the generation of artefacts (light-generated halo around the cell body) due to refraction anomalies, which could hide important morphological information (Debeir et al., 2008). Doing this will enhance contrast, allowing for a better separation of the cells from the background making tracking analysis easier (Meijering et al., 2008). When selecting sites for tracking, it is generally best to pick at least two fields from the same condition; this ensures enough cells to track, and gives a technical duplicate for that condition. If tracking a fluorescently tagged cell, it is important to try for and compensate for illumination bleaches and phototoxicity. The generation of free radicals is not only toxic for cells but can also alter their behaviour. Some methods have been proposed to deal with this problem such as minimising the signal-to-noise ratio (SNR), but in reality this function should be maximised for optimal tracking (Meijering et al., 2012). Sophisticated filters have been developed to generate pseudo-fluorescent images, but these are quite expensive and not widely adopted in many labs (Meijering et al., 2012). Also, the requirement for shutter control is important in these experiments so that only the area of interest is illuminated at the time of capture. Once the correct settings have been assigned and the setup is complete automated image acquisition can begin.

7.3 OVERVIEW OF AUTOMATED AND MANUAL CELL TRACKING SOFTWARE

Upon completion of a cell migration experiment, the next step is to reconstruct each cell trajectory from the sum of all the frames in a sequence. The file type for tracking will depend on the software being used. Typically movies will need to be saved as either.tiff or .avi to be able to be read by most tracking software. Currently, there are two main approaches to cell tracking: automatic and manual. We describe a selection of software available using both these methodological approaches in order to gain an understanding of the requirements of their use.

7.3.1 *Automatic Tracking*

With the advance of modern computers, storage and RAM capacities, automated cell tracking systems are now a promising method for determining the migratory potential of cells in 2D time-lapse videos. This type of software operates to fulfil a set of requirements in order to effectively track live cells. Firstly, some degree of image pre-processing needs to be done in order to ensure a good contrast between the inner part of the cells and their surrounding region. Next is a segmentation step which separates the cells from the background, allowing an object to be recognised. Segmentation algorithms have not been perfected yet and so sometimes post-processing stages are performed in order to give a more accurate identification of cells in the field of view. Theoretically, once these steps have been performed quantitative measurements can be extrapolated from the movies without further interference (Debeir et al., 2008; Meijering et al., 2012).

When it comes to automatic tracking software, the majority of tools assume that cells are easily differentiable from their immediate surrounding. This is usually envisaged as cells modelled as bright regions against a darker background. However, whether it be a phase, Nomarski or fluorescent microscopy scenario, differences in pixel intensities are not always obvious. Most of the available tracking software takes this into account and utilises suitable filters to deal with a non-uniformed signal-to-noise ratio (Meijering et al., 2012; Peskin et al., 2011). Examples of different pre-processing methods are seen in three tracking software packages, DYNAMIK, iTrack4U and ADAPT, all of which are free to use (Barry et al., 2015; Cordlières et al., 2013; Jaeger et al., 2009). DYNAMIK uses median filtering, which works by evaluating the brightness of each pixel and then comparing this with the value of its neighbouring pixel. This then generates a medium value in a so-called pixel neighbourhood and replaces any pixels with values that vastly deviate (Jaeger et al., 2009). iTrack4U uses a histogram filter that finds an estimate of the surrounding grey values and then uniformly redistributes these intensity values to give the most accurate signal (Cordlières et al., 2013). ADAPT, a plugin for the open-source program ImageJ uses Gaussian filtering, which functions by smoothing and blurring the image to remove noise and then re-represents each pixel as a mean of its surrounding pixels (Barry et al., 2015). While all these filters do the job of improving the images in their own particular way it is often the case that inexperienced or inadequately trained researchers do not fully understand the consequences of their use. For instance, Gaussian filtering is quite commonly used and works quickly and intuitively. The fact that it is computationally efficient gives it an advantage over median filtering, which has high computational cost. However, median filtering is much better at preserving cell edge detail, which is reduced in Gaussian filtering (Arias-Castro and Donoho, 2009). Histogram filtering is also computationally efficient and greatly improves the contrast of images, but the drawback of this filtering method is that it is indiscriminate, and thus can increase the contrast of background noise and reduce the usable signal (Chuang, 2001).

Once pre-processing has finished, the image needs to be divided into one region of pixels that is identified as cells and another region of pixels that is associated with the background. This process is known as segmentation and, as with the filtering step, there are multiple methods used in order to segment images (Xiong and Iglesias, 2010). Two main approaches exist for object segmentation: pixel detection and model-evolution. The concept of pixel detection is relatively simple whereby the program simply segments based on selecting pixels whose intensity is above a predefined threshold. It works well when the objects are well-separated (i.e fluorescent imaging), but for non-fluorescent images, such as phase contrast, adjusting the threshold to capture darker or lighter cell boundaries is needed (Xiong and Iglesias, 2010). Model-evolution uses more elaborate mathematical algorithms than grey-level detection and encompasses a larger spectrum of techniques with the two most adopted algorithms for cell tracking being the snake model and the mean-shift model (Li et al., 2006). The snake model works by fitting an active contour around the cell, which evolves over time according to the limits generated by the convexity and curvature of the cell shape in addition to the boundary intensity (Sacan et al., 2008). The mean-shift model approximates the cell border of a cell as a polygon, referred to as a region or kernel. Each kernel is divided into sectors which nest two isosceles triangles, one being sensitive to light pixels and the other being sensitive to dark pixels. The tips of all these triangles correspond to the centre of the cell and are displaced according to the adjustment of the cell centroid in each frame (Cordlières et al., 2013). Several automatic tracking programs use the pixel detection method including CellTracker, DYNAMIK and ADAPT. While Icy and CellTrack use the snake model, and iTrack4U uses the mean-shift model. The pixel detection method is useful in that it is not computationally expensive, and cells can be easily segmented. However, unless cells are fluorescently labelled, this technique may not be applicable as in most cases a big enough difference in pixel intensity between the objects of interest and the background might not exist (Möller et al., 2014). The mean-shift and snake models are both more capable algorithms for detecting and tracking migrating cells under phase-contrast microscopy. The mean-shift algorithm would benefit those who are tracking cells with very dynamic shape changes as it makes assumptions for cell speed based on the information retrieved from the displacement of internal kernels (Cordlières et al., 2013). However, the process is highly sensitive, meaning that any modification in cell shape may be considered movement and thus generate exaggerated cell speed (Cordlières et al., 2013). The snake model is advantageous in that it can locate object boundaries automatically from the initial contour and give a linear determination of the object shape at the convergence time without the need for extra processing, making it more computationally efficient compared to the mean-shift model. However, the technique is dependent on a strong image gradient in order to get optimal segmentation (Möller et al., 2014). The major drawback of both these model-evolution techniques is the ability to handle topology changes of the contour, which means that if a cell divides, or if two cells' contours make contact, then tracking may fail (Li et al., 2006). In reality, topology preservation is a problem for most automated tracking software and is currently a major concern. Pixel-detection software such as DYNAMIK tends to undersegment, meaning that two adjacent cells may get wrongly connected, while other tracking software such as CellTrack oversegments meaning that cells may get split into a number of parts (Jaeger et al., 2009).

Some software packages utilise post-processing software which complements image segmentation in order to better identify objects of interest. An example of this is DYNAMIK, which uses the Sobel operator to remove any inclusions within the cell, which allows the program to better define the cell boundary. Again, while advantageous in order to get the most accurate measurements during cell tracking, these operators do increase the computational burden (Jaeger et al., 2009).

After processing has been completed, tracking of selected cells in movies can be performed. The software is now tasked with having to link a cell from one frame to the next frame without losing essential information. Several linking strategies currently exist but the most robustly used in automated cell tracking software is the nearest-neighbour and deformable model approaches (Meijering et al., 2012). Nearest-neighbour is used by several tracking programs including DYNAMIK and ADAPT and works by associating each segmented cell in one frame with the nearest cell in a subsequent frame. This association can be based on information such as spatial distance, boundary points, centroid position, cell size, orientation and curvature. Essentially nearest-neighbour is an efficient and simple protocol for generating cell trajectories, but mismatching can occur if dealing with rapid cell movements and a high number of cells. The more information that is used the less ambiguity there is between cells, but it then becomes computationally taxing, limiting the overall number of cells that can be tracked (Meijering et al., 2008). Deformable models used by programs such as Icy, CellTrack and ITrack4U allow for more complex tracking. The principle of deformable models is that segmentation is applied in all frames of the movie, meaning that contour, curvature and centroid position information are used as a basis to correctly identify the cell in each subsequent frame. This technique works well, but if the cell displacements are more than the diameter of that respective cell in the next frame, matching will be lost and more sophisticated algorithms are required in addition (Meijering et al., 2008). Alternatively, shorter time intervals between image capture can be used to reduce this possibility.

The final step is the extraction of data from the software to generate meaningful results. Some tracking software including CellTrack, icy and iTrack4U require only windows as a platform to operate and display information. Others such as ADAPT, are Java-based plugins that need to be booted up in ImageJ in order to function. DYNAMIK on the other hand has been written and needs to be read in MATLAB, which is proprietary software that may not be freely available to all cell biologists (Meijering et al., 2012). All these software packages compute essential information gathered from the x, y and z coordinates of cell trajectories. With the exception of ADAPT, all the automatic programs give information about cell speed over time. ADAPT works out the velocity, which is a relative measurement of the motility of a cell, but instead of measuring speed alone, it factors in direction of travel, and so it is more accurately a measurement of the rate of change. All programs give some form of information about cell shape, including roundness, area or perimeter. DYNAMIK and ADAPT record directionality, which is more useful for chemotaxis assays such as Dunn chamber assays (Barry et al., 2015; Jaeger et al., 2009). iTrack4U measures persistence quite thoroughly, which is useful if one hypothesis predicts that cell polarity or the randomness of migration paths might be affected in their experiment

(Cordlières et al., 2013). Programs that use deformable models such as Icy and CellTrack also provide additional information on contour deformation of the cell over time (Meijering et al., 2012).

7.3.2 *Manual Tracking*

Manual tracking is still regarded as the gold standard of measuring cell movement across a planar 2D surface. It is the most accurate and commonly used approach to obtaining data from a cell trajectory over time (Meijering et al., 2012). Unlike automated cell tracking, manual tracking software is relatively straightforward to use, and all programs work on a similar principle, which is to choose a target cell in one frame, and then connect this object to the same physical entity in sequential frames (Meijering et al., 2012). This point-and-click system records x and y coordinates which compute into a given number for the horizontal and vertical location of that pixel (Cordlières, 2005). The two most well-known and used manual tracking programs are Manual Tracking and MTrackJ, both of which are Java-based plugins for ImageJ (Meijering et al., 2012). Manual Tracking was developed by Fabrice P. Cordlières and works by inputting key information (described in more detail later) about the setup of your movie, which allows the plugin to retrieve data. This information is displayed in a table which gives the user xy and xyz coordinates, as well as slice number, distance, velocity and intensity of the selected pixel of volume (Huth et al., 2010). The Manual Tracking plugin also features a 'centring correction' option which allows users to track the centroid of a cell more accurately. This feature works by adjusting your tracking according to the bright and dark pixels that are directly in the search square around your click. While this feature can prove useful when tracking multiple cells over long periods of time, its efficiency for accurately modifying tracks is dependent on pixels within a frame being either very bright (local maximum) or very dark (local minimum) (Cordlières, 2005). MTrackJ was developed by Erik Meijering and his group to complement automated cell tracking when movies were not suitable to be tracked in this way (Meijering et al., 2012). Similar to Manual Tracking, MTrackJ opens up in ImageJ as a window with several options including those to track cells. The program works in a similar manner, in that you open up your movie and then utilise a point-and-click system to measure the trajectories of your cells of interest. The setup of MTrackJ is slightly more complicated than that of the Manual Tracking plugin in that the latter directs you to a screen where you input all the necessary parameters, while in the former you have to browse over a few menus. For instance, unless reading the online manual for MTrackJ, someone unacquainted with the software may forget to calibrate the movie's pixels correctly using the *Image > properties* tab and could then end up with the incorrect spatial unit readings for their trajectories. Once you are familiar with the software though, MTrackJ is convenient and easy to use for cell tracking. The tab *Add* must be clicked in order to start tracking, and once finished the user clicks the tab *Measure*, which opens a window similar to that for manual tracking, which displays an array of information about your cell's tracks. The information in the window includes track number, slice, xyz coordinates, time, length of track from start to finish, D2S, D2R and D2P units, which can be used to measure Euclidean distance and directionality, velocity and angle of trajectory. These metrics taken together can describe the cell's migratory behaviour over time (Meijering, 2006). The outputs of both Manual Tracking and MTrackJ can be put into other third-party support software such as the chemotaxis and migration tool, a standalone or plugin for ImageJ developed by Ibidi®, or DiPer™, an open-source computer program developed by Gorelik and Gautreau (Gorelik and Gautreau, 2014). These programs allow you to generate additional information from your data, including forward migration index, directionality, speed and migration track or rose plots. Manual tracking data can also be placed in the notebooks in Mathematica developed by Graham Dunn, Daniel Zicha and Gareth Jones (described in more detail later) which generates similar information to the other programs with the addition of mean speed and persistence over time, and including statistical analysis within and between data sets. In addition to these two programs, some automated cell tracking programs such as Icy also have the feature to manually track cells. However, this feature is generally suboptimal in these programs and thus it is more common for researchers to use Manual Tracking or MTrackJ, which are well established and robustly used (Meijering et al., 2012).

7.3.3 *Comparison Between Automated and Manual Tracking*

There are no comprehensive studies directly testing the differences between manual and automated cell tracking programs, but there are several reviews from researchers on their experiences of using both kinds of software (Huth et al., 2010; Meijering et al., 2012). Here, we will give an overview of the major advantages and disadvantages associated with manual and automated tracking software.

Manually tracking cells is still common practice among researchers studying motility (Meijering et al., 2012). Typically, all software associated with this type of tracking method, whether it be the Manual Tracking plugin or the MTrackJ plugin, is user-friendly. Online instructions provide enough detail on the fundamentals of using these programs, and the simplistic interface and straightforward output means that labs unfamiliar with cell tracking can begin with relative ease. While there are instructions available, this user-friendly aspect does not remain entirely true for the majority of automated cell tracking programs. These systems generally come with overwhelmingly detailed – or not detailed enough – instructions which could deter users from using automated tracking programs with a slightly perplexing interface.

When tracking cells manually, the data is arguably more comprehensible to the observer as they may be more likely to notice oddities or differences in cell shape, behaviour, proliferation or death. Some automated tracking programs such as ADAPT do provide a nice feature which can give information about cellular dynamics during the movie, but the accuracy of this information is heavily dependent on the signal-to-noise ratio and segmentation process, which has been shown to vary (Möller et al., 2014). In automated cell tracking, the processing and segmentation steps are fairly onerous and time-consuming for the researcher, taking up to 2 hours, even when familiar with the software (Cordlières et al., 2013). Ideally, using a fluorochrome to tag cells is the best option for this system as it would allow for quick processing and more accurate segmentation. However, this can lead to phototoxicity issues and unusual behaviour of the cells (Meijering et al., 2012). Precise identification of non-fluorescent cells still remains a challenging task in low contrast regions, and while deformable segmentation models remain promising, they require long processing times, making them unsuitable for high throughput data generation (Huth et al., 2010). As we discussed earlier, different automated programs offer different filters and algorithms, none of which has been perfected yet. Over- and undersegmentation remains a problem, which has been well noted, with one study in particular showing a regional dirt particle being associated with a cell by the program, giving a larger variance during tracking (Möller et al., 2014). Another major issue is cells entering into and leaving a frame. In manual tracking, it is easy for the observer to simply stop tracking, click the end track function, and then restart by tracking a new cell. In automated cell tracking, falsely matching cells entering and leaving the field of vision with a new, but different, cell is possible. This is due to the thresholding algorithm reducing the contour to a single pixel at the edge of the frame, but not entirely eliminating it (Jaeger et al., 2009; Möller et al., 2014). Automated cell tracking is also prone to errors in the identification of the cell in every frame of a sequence, and then linking this same cell as one physical entity in the entire image sequence. Since these two procedures are usually independent of each other, the lack of feedback between them makes the tracking method prone to failure (Kalaidzidis, 2009). This leads to major problem of erroneous deletion of cells tracks, due to missing cell-to-cell associations. This deletion of tracks has been reported to lead to large errors in the data as all the mean displacement values beyond the time point when the track was deleted are now lost (Huth et al., 2010). Generally, to reduce the risk of this occurring, it is recommended to take more frames over the total time-course of the movie so that the cell never displaces beyond its own cell diameter from one frame to the next (Meijering et al., 2012). While it is easily possible to take more frames, in a video with many cells, this is likely to increase the computational burden and would require some knowledge of how fast the cells move. Many programmers are hard at work refining the algorithms to better deal with cell proliferation and movement into and out of a frame, so hopefully updates to the systems will appear in the near future. Another drawback to many automated tracking systems is that they offer little or no functionality to manually inspect cell trajectories (Meijering et al., 2012). Correction during the course of tracking could reduce segmentation errors, while manually tracking, or setting a mean manual trajectory as a control for the system could be a nice alternative to correcting the displacement values (Meijering et al., 2012).

Even though the setup time of automatic cell tracking is long, once it is complete the overall tracking can be done quite quickly. Unfortunately, this is not true for manual tracking, which is time-consuming and labour intensive. Indeed, studies suggest that you can track a higher number of cells in a reduced timespan with automated tracking (Cordlières et al., 2013). This leads to the problem that if the observer has a large number of cells on the screen, generally only a subset of the cell population is selected to represent the whole population. Thus variables such as cell behaviour and true cell speed could differ significantly between two different experimenters working on the same cell tracking experiment (Huth et al., 2010). In addition, while the technique is straightforward, gaining good results is highly dependent on the operator's skills and perception. For example, manual tracking requires the operator to identify the cell centroid consistently across all frames, to reduce variability of the data (Huth et al., 2010). Since automated cell tracking uses centroid determination by pixel and edge detection, this is rarely a problem and thus more likely to produce reproducible results. Manual tracking is also more likely to favour biasing, since the user essentially re-watches the movie with each cell they track. From this, a priori knowledge is accumulated and this could unconsciously make the observer favourably select cells from within the population as the eye is good at selecting faster moving objects. This remains especially an issue when dealing with cells of a heterogeneous nature, and so essential information such as speed and shape may be lost (Möller et al., 2014).

7.4 INSTRUCTIONS FOR USING IMAGEJ TRACKING

ImageJ is a public domain image processing program developed by the National Institutes of Health (Collins, 2007; Schneider et al., 2012). It is an open architecture program that only requires an operating system which offers a Java runtime environment (Collins, 2007). This makes it a popular and well-used program because the majority of operating systems, including Windows, Macintosh and Linux, are compatible with Java-based applications. Being open-source software, ImageJ allows users to create modules, commonly referred to as plugins, which are then made freely available for others to use for their intended research purpose. Once put into practice, these plugins can then be improved with regards to their original function, and altered to gain new functions in research. ImageJ currently offers over 400 plugins, making it a great tool in the field of medical imaging and microscopy. These plugins allow the program to be as simple or comprehensive as the user requires, and due to its high usage in a variety of fields, the possible amalgamation of the program means third-party software compatibility is unlikely to be an issue (Collins, 2007). Fiji (www.Fiji.sc) is newer variant of ImageJ, including support for 64-bit operating systems and newer plugins (Schindelin et al., 2012). This can be used instead of ImageJ.

After image acquisition of your experiment has been completed it is good to set some parameters before performing tracking analysis in ImageJ. For instance, if you have a population of cells that are actively proliferating, the assigned length of time tracking these cells needs to be adjusted against the total time of the movie. For example, if the total time of the movie is 16 hours, and the cell divides at 14 hours, then tracking should be preferentially stopped at 13 hours. A different rule applies to cells that undergo apoptosis. It is normal for some cells to undergo cell death during the course of the movie. These cells can be tracked until they start to round up, at which point tracking should

Figure 7.1 Display window for ImageJ running on Windows 7 Professional.

be stopped. If you notice an unusually high number of cells undergoing apoptosis then this video should not be tracked as it is likely that culture conditions have affected the experiment. You may also get the likely occurrence of cells going out of the field of view and new cells entering the field of view. If this is the case it is generally acceptable to track a cell exiting the field until it is no longer visible, but it is not acceptable to track a cell that is not present on screen in the first frame of the movie.

Upon opening, ImageJ presents itself as a relatively small window (Figure 7.1). The running of this window uses little of the system memory, but it runs very quickly, rarely ever displaying any slowdowns or crashing. It offers a user-friendly interface with a variety of functions already available for image processing.

The display window of ImageJ can be broken down into three main parts, two offering functionality, and one displaying information, marked 1, 2 and 3 in the figure:

1. Functional tabs

File: Here you can open, save and print images. ImageJ has the ability to load and support 8-bit, 16-bit, 32-bit and 64-bit images that exist in a variety of different file formats (.jpeg, .tiff, .gif, .bmp, .png, .dicom, .fits and .avi).

Edit: Here you can edit images. This can include distorting size, and adding text information and drawings.

Image: This allows the user to modify and covert images (8-bit, 16-bit, 32-bit, 64-bit, RGB), as well as carrying out geometric operations if several images taken from different light path channels need to be stacked.

Process: Processing of images can be done here. A series of filters (Gaussian, median, salt and pepper etc.), edge detection and arithmetic operators are available.

Analyse: Here you set the parameters for your images and perform statistical measurements, add scale bars and make histograms.

Plugins: Accessible menu to utilities and plugins. The list of plugins available is dependent on those that have been downloaded and placed into the plugins folder of ImageJ, found in the programs folder of the C: drive.

Windows and Help: These allow organisation of tabs found within the window and give some instructions on using ImageJ, respectively.

2. *Tools*

On this row of the ImageJ window, quick point-and-click tools can be found that are useful to the user during image processing. A series of shapes are available to draw for cut-and-paste modification. A free-draw tool exists to allow the user to define boundaries. Line and angle tools can be useful if looking at specific measurements within a frame. Text, colour and shape art can be found, as well as magnification and click-and-drag tools.

3. *Diagnostics window*

Here, basic information is displayed. When no images are open, it will give information about the version of ImageJ being used. If an image has been opened, it will provide details such as the x, y and pixel density value.

- To begin tracking in ImageJ, the Manual Tracking plugin which should sit in the *Plugins* menu needs to be booted up. Once this is done, the window in Figure 7.2 should present itself on the main screen.

- Next, either drag your time-lapse movie into the ImageJ window, or open it conventionally in ImageJ using the file window. Movies should be saved as either .tiff or .avi for tracking. If opening a movie saved as .avi, the AVI reader window might appear. If this occurs make sure the first and last frame numbers are correct, and if so then click ok and continue.

- Before tracking, the parameters of the movie need to be set in order to retrieve correct pixel and position information. This information can be set in the option boxes shown in Figure 7.2(6). In this example, a picture has been taken every 5 minutes, as indicated by the information in the *Time interval* boxes. The x/y calibration value also needs to be set. This is done by imaging a stage graticule which will give a defined number of pixels per micrometre in x and y; then this number is divided by the magnification to give the final value, which in our example is 1.37 μm. Finally, enter the z calibration value, which in our example is 1 because each stack exists on a single plane.

- The centring correction option (Figure 7.2(2)) can help the user track more accurately by automatically adjusting the coordinates of each pixel according to the local maximum, the local minimum or the barycentre of intensity in the surrounding of a cell. If choosing to use this option the *Search square size* box in Figure 7.2(6) will need to be changed to a user-defined square size measured in pixels.

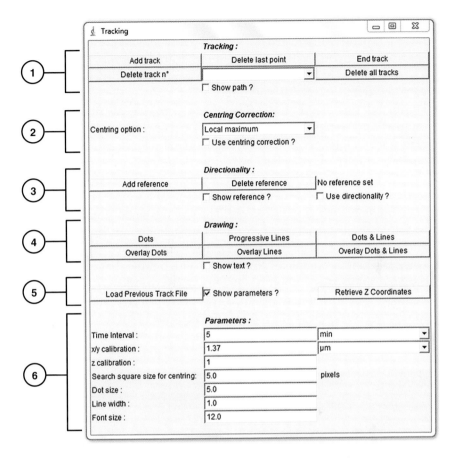

Figure 7.2 Interface for manual tracking plugin opened in ImageJ.

- A directionality option also features in Manual Tracking, as seen in Figure 7.2(3). This method relies on the user selecting a reference pixel point in the frame, which can be seen as a circular region of interest. This is used as a determination of the first direction of movement with further directions of each cell being compared to this reference during tracking. Direction changes are defined by changes between three successive points passing a 90 °C limit.

- To start tracking, simply click *Add track* as seen in Figure 7.2(1). This will initiate the tracking protocol of the program. Once the user points and clicks on a cell of interest a new window will appear which contains all the results from the tracking. If the cell undergoes apoptosis or proliferation you can click on the *End track* tab and then restart cell tracking by clicking the *Add track* tab again. If you make a mistake or accidently lose track of the cell being tracked you can click on the *Delete last point* and *Delete track n°* tabs (make sure to select the specific track you want to delete in the drop-down tab) respectively. There is also a *Show path?* box which can be ticked if you want to view the cell trajectory as you track.

- During cell tracking several different views can be generated to display the tracking results as seen in Figure 7.3. This can be done with the tabs under *Drawing* as seen in Figure 7.2(4). The options available here are to display your results as dots or progressive lines both in a simple format where the dots and lines appear on a black background, or overlaid over each frame of your movie. An option to overlay both dots and lines together also exists. The size of the dots and the width of the lines can be changed in the parameter option boxes *Dot size* and *Line width* as seen in Figure 7.2(6).

- Once tracking is complete, you are left with a table that has recorded essential information about each cell's trajectory including the track number, xy coordinates, distance travelled, velocity and the corresponding pixel value (Figure 7.4). The first time point for each track is

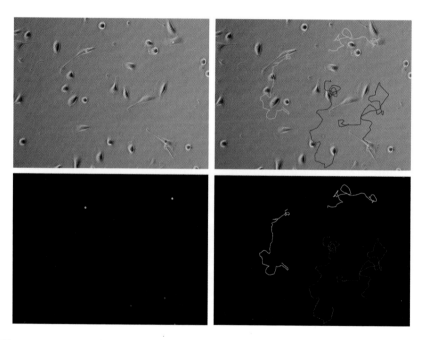

Figure 7.3   Example of the drawing features available in manual tracking to allow the user to identify which cells have been tracked.

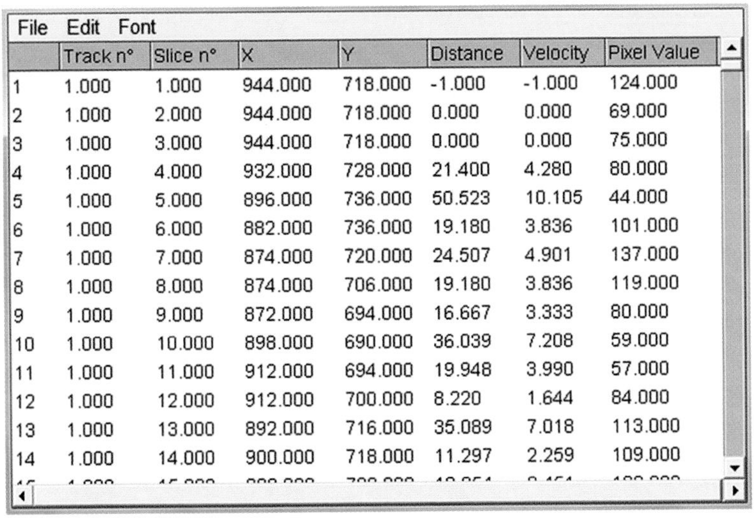

	File	Edit	Font					
	Track n°	Slice n°	X	Y	Distance	Velocity	Pixel Value	
1	1.000	1.000	944.000	718.000	-1.000	-1.000	124.000	
2	1.000	2.000	944.000	718.000	0.000	0.000	69.000	
3	1.000	3.000	944.000	718.000	0.000	0.000	75.000	
4	1.000	4.000	932.000	728.000	21.400	4.280	80.000	
5	1.000	5.000	896.000	736.000	50.523	10.105	44.000	
6	1.000	6.000	882.000	736.000	19.180	3.836	101.000	
7	1.000	7.000	874.000	720.000	24.507	4.901	137.000	
8	1.000	8.000	874.000	706.000	19.180	3.836	119.000	
9	1.000	9.000	872.000	694.000	16.667	3.333	80.000	
10	1.000	10.000	898.000	690.000	36.039	7.208	59.000	
11	1.000	11.000	912.000	694.000	19.948	3.990	57.000	
12	1.000	12.000	912.000	700.000	8.220	1.644	84.000	
13	1.000	13.000	892.000	716.000	35.089	7.018	113.000	
14	1.000	14.000	900.000	718.000	11.297	2.259	109.000	

Figure 7.4 Window displaying the results of cell trajectories gained during tracking using the Manual Tracking plugin.

recorded as –1, as the distance and velocity cannot be calculated. Note that the unit for velocity and distance depends on the unit selected for the time and xy calibration.

- Once it is complete, you can save your results as an Excel file.

7.5 POST-TRACKING ANALYSIS USING THE DUNN MATHEMATICA SOFTWARE

After tracking has been completed in ImageJ, the generation of standard cell motility data can be done using other plugins such as the chemotaxis tool. However, more complex post-tracking analysis can be done using programs such as Mathematica, which is an advanced computational software program conceived by Stephen Wolfram and developed by Wolfram Research of Champaign, Illinois (Maeder, 2014). Mathematica uses the Wolfram language as its programming language, which has immediate built-in access to extensive scientific and technical data (Maeder, 2014). Professor Graham Dunn and his colleagues have written in-house software that reads into Mathematica, called the chemotaxis notebook. This notebook uses the x and y values from the Manual Tracking output to find the true cell speed and persistence, as well as to generate graphical and statistical data.

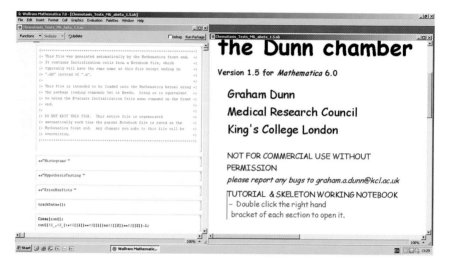

Figure 7.5   Wolfram Mathematica 7.0 interface displaying the chemotaxis notebook program.

Detailed instructions and the logistics of each module have been described previously by Zicha et al. (1997).

The notebooks can be obtained on request by contacting Professor Gareth E. Jones: gareth.jones@kcl.ac.uk

- To begin the analysis, Wolfram Mathematica 7 needs to be booted up on the desktop.
- Once this is done, the user will need to load the chemotaxis notebook into Mathematica. To do this click *File* > *Open*, then select both notebook files, .nb and .m.
- When it is open, you should be presented with two main windows as seen in Figure 7.5. The window on the right is called the kernel, and it interprets syntax to generate usable applications. The window on the left is known as the front end, which provides an editable graphical user interface for modifying code.
- When opening the notebook in Mathematica, the user is greeted with 6 fields that can be opened to perform different types of analysis (Figure 7.6).
- The first field provides a brief introduction to the chemotaxis notebook and detailed instructions on the Dunn chamber assay, a tool used for studying the chemotactic potential of cells. If the user has performed this experiment, they may wish to read here how the details of the assay relay to analysis that can be done in this notebook.

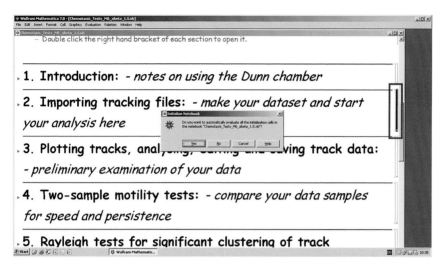

Figure 7.6   Initiation message that appears over the chemotaxis notebook when the user presses Shift Enter.

- To begin the analysis of your tracking data in Mathematica, the user first needs to import their file into the notebook. To do this, the rightmost bracket in the right corner of the cell (boxed in red in Figure 7.6) needs to be activated. Click on this bracket and then press Shift Enter. The first time you do this it will prompt a message asking if you want to automatically evaluate all initialisation cells. Click *Yes* and then double-click on the rightmost bracket, which should open up the cell to reveal the Track Importer (Figure 7.7). This process must be repeated when opening all subsequent cells.
- In section 1 of the Track Importer you have a choice of three programs (Andor, ImageJ, Metamorph) from which to import your tracks (Figure 7.7). If selecting ImageJ, be aware that the file needs to be saved as text tab-delimited, and will only work if the cells were tracked in the Manual Tracking plugin as opposed to the MTrackJ plugin.
- In section 2 of the Track Importer, import your file from its current destination.
- Since Track Importer does not import any time data or scale factors, section 3 requires you to input the parameters of your movie similar to what was done in the Manual Tracking plugin. In our example seen in Figure 7.7, we have entered the same values that were entered in the Manual Tracking plugin, 1.37 for the x and y coordinates, and 5 minutes for the time interval. If you are looking at cell movement in response to a chemoattractant gradient, and the direction of your data

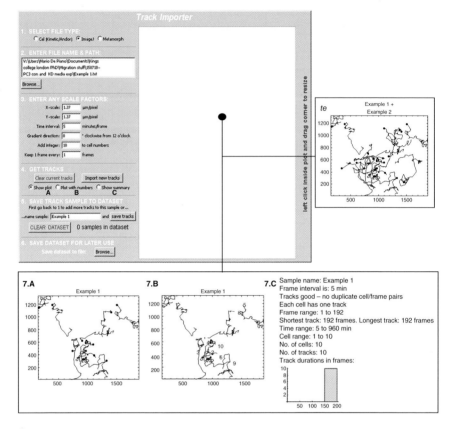

Figure 7.7   Importing tracking files into Mathematica.

is not vertically upwards, then you can rotate this data in the *Gradient direction* field. If importing multiple files, and you need to keep the original cell numbers, you can change the integer in the *Add integer* field to 0.

- Once content, you can click the *Import new tracks* tab in section 4 of the Track Importer.
- Once it is imported, section 4 gives you three options on how to view the data: as a regular plot as seen in 7.7.A, as a plot with number as seen in 7.7.B, or as a summary of the data as seen in 7.7.C. The summary in 7.7.C provides good information on your tracking data, indicating whether you have accidently tracked the same cell twice or have an anomaly in your population which may skew the data.
- If you want to add a new data set, you can click on the *Clear dataset* tab in section 4, which should clear the current tracks, and then you can import a new data set. If you want to add a set of tracks to your

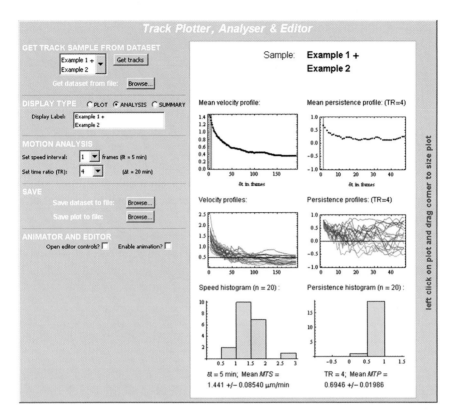

Figure 7.8 Track plotting, analysing and editing window as displayed in Mathematica.

current data set then you simply browse the file and click *Import new track* instead of *Clear dataset*.

- Once your data is imported, you can save the data set in section 5 of the Track Importer. Be aware that you need to do this every time you import your tracks, otherwise when you add new data sets, the original ones will not be stored for use in further analysis.
- In the drop-down menu under the *Get track sample from dataset* menu in the *Track Plotter, Analyser and Editor* window click on your saved dataset (Figure 7.8).
- In the *Display* menu the *Plot* and *Summary* option will give you the same information as seen in Track Importer.
- When selecting the *Analysis* option of the *Display* menu the user is presented with the results as seen in the main window of the *Track Plotter, Analyser and Editor* cell (Figure 7.8). Here, you are presented with a series of graphs that give information about the speed/velocity and persistence of your tracks. The top two graphs labelled *Mean*

velocity profile and *Mean persistence profile* show the mean velocity and persistence respectively of all cell tracks that were imported. The individual velocity and persistence of your cells can be seen directly below these two graphs. The two bottom graphs in Figure 7.8 show histograms for the mean cell speed and persistence in addition to their corresponding standard error of the mean. These histograms also provide a good indication of the variability in speed and persistence that you have within your cell population.

- The *Track Plotter, Analyser and Editor* window also contains a *Motion analysis* menu which can be used to alter the time ratio and speed interval if the user wishes to look at differences in cell movement at a specified interval.
- In the *Animator and editor* window, the user has the ability to change the graphics of the tables as they see fit by ticking the *Open editor controls* and *Enable animation* tabs.
- The chemotaxis notebooks in Mathematica are also capable of performing statistical operations on your data.
- In Figure 7.9, you have the display window for the *Two Sample Tester* which allows the user to perform a simple t-test on their cell track data.
- In the *Get two track samples from dataset* menu there are two dropdown boxes where you can select your datasets and then click the *Get tracks* option to import them into this window. There is also a *Browse* tab where you can import previously used data.

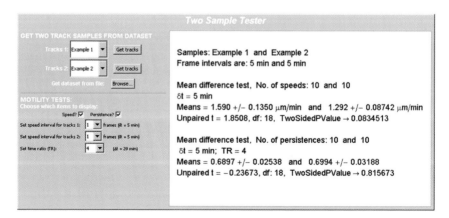

Figure 7.9 Two sample t-test window as displayed in the Mathematica chemotaxis notebooks.

- Then in the *Motility tests* menu, select *Speed* and *Persistence*, if you want to do statistics on both these processes, and the statistical information should be displayed in the right side of the window.
- The information presented here will give an indication of the time interval between frames, speed and persistence for each set of tracks, and the t-test. The notebook gives both the unpaired t-test value and the two-sided P-value. The unpaired t-test value tests whether there is a difference between the two cell populations in one tail of your distribution and, if below 0.05, will be significant. The two-sided P-value provides slightly more stringent analysis by comparing both ends of your population meaning that 0.025 is in each tail of your distribution.
- The Rayleigh test is another statistical test that can be used in the Mathematica chemotaxis notebook (Figure 7.10). It functions to test the uniformity of a sample of cells moving in a particular direction.
- As before, import your tracks from the drop-down menu under *Get track sample from dataset.*

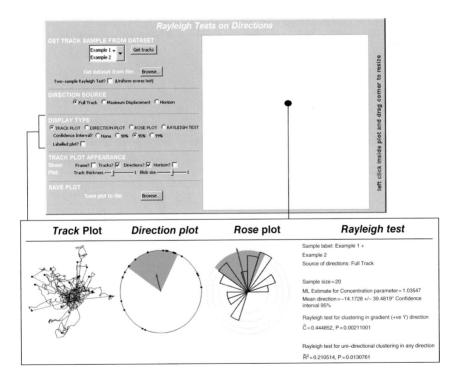

Figure 7.10   Rayleigh test of *Directions* window as displayed in the Mathematica chemotaxis notebooks.

- Below is a *Direction source* option. This gives you three options on the information you want to include in the statistical test from your dataset. The *Full track* option takes the full direction of your track, start point to end point, into account. However, since some cells can back track, there is a concern that you may be losing information of the total distance travelled by each cell. One solution to this is to use the *Maximum displacement* option, which takes the direction from the start point to the furthest point the cell has travelled. Another option available is *Horizon*, which lets the user define a fixed start and end point. This allows the exclusions of tracks that don't travel very far from the centre of mass and so could mask a true gradient effect.
- Next, under the menu *Display type*, you can choose to view that data as a track, which has blue lines protruding out of the centre of mass to show whether the full track, maximum displacement or horizon points are being measured. You can also view the data as a direction or rose plot, which both have an arrow and highlighted green area navigating towards the direction that the cells are moving. The rose plot also includes circular histograms of the frequency of tracks moving in towards certain directions. The green area in both these plots will only appear if there is significant clustering towards a direction and can be made more stringent by selecting a higher confidence interval.
- You are also given a summary of the Rayleigh test results, which gives you a P-value for cells clustering towards a gradient, and a P-value for cells clustering in a unidirectional manner in any other direction.
- Once the analysis is done, as with the previous cells, there is an option to modify the appearances of your graphs under the *Track plot appearance* menu.
- The Mathematica chemotaxis notebook also allows the user to perform the Moore test (Figure 7.11). The Moore test allows for more elaborated analysis about the direction your sample of cells is travelling. For instance, the Rayleigh test tells us which direction the cells are clustering towards, but the Moore test gives information about the ones that travel up-gradient, which typically have longer tracks.
- As with the previous tests, import your data and click *Get tracks*.
- Once this is done, you are given the option under the *Display type* menu to view your data as a track plot, vector plot or Moore test.
- As shown in our example in Figure 7.11, just like the Rayleigh test, you are given the option to assimilate this data from the full track or the maximum displacement values. Again, this will depend on the research question you are trying to answer.

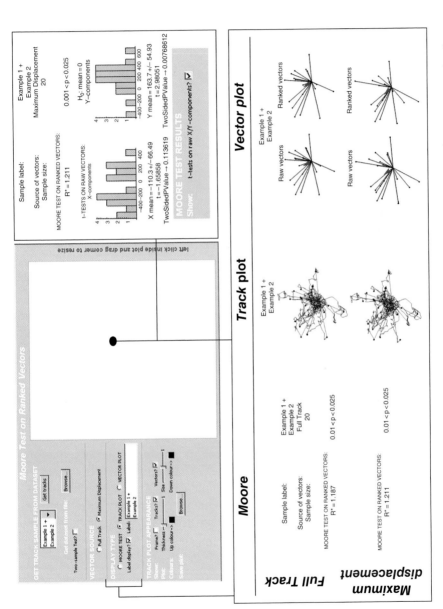

Figure 7.11   Moore test on ranked vectors window as displayed in the Mathematica chemotaxis notebooks.

- The Moore test statistic R* is worked out by the length divided by the square root of your total number. In our example, it is not significant.
- Finally, the Moore test also allows you to do t-tests on vector x and y mean components. You can use this for the y-components, to test if the response of directional migration coincides with the known gradient axis. If it does, the y mean value should significantly differ from 0, while the x mean should not.

7.6 SUMMARY AND FUTURE DIRECTION

In this chapter we have discussed in some detail the positive and negative aspects of manual and automated cell tracking. While both systems have their limitations, they are more than sufficient for measuring the movement and direction of cells across a planar surface. Models developed to accurately measure cell migration in 2D environments have been the main focus over the past 20 years. However, as our understanding of both *in vivo* embryonic cell migration, leukocyte trafficking and cancer metastasis becomes more comprehensive there is an increasing interest in applying this knowledge to the study of cell motility in 3D. Studies in 2D give us a fair insight to the mechanisms of cell-substrate interactions during movement, but these studies cannot recapitulate the complex viscoelastic matrix characteristics of *in vivo* migration. Currently the main problem with studying cell motility in 3D is the lack of quantitative mathematical models to effectively generate high quality data. Most analysis is done manually and subjectively through counting or thresholding staining intensities. These methods are likely to miss subtle changes between two groups and can only deliver basic information about the movement of cells through a 3D matrix. Since the focus is shifting to study motility in the third dimension, it is likely that in the coming years we will see an increased demand for tracking software that can generate high quality quantitative data from 3D assays.

7.7 REFERENCES

Arias-Castro, E. and D.L. Donoho. 2009. Does median filtering truly preserve edges better than linear filtering? pp. 1172–1206.

Barry, D.J., C.H. Durkin, J.V. Abella and M. Way. 2015. Open source software for quantification of cell migration, protrusions, and fluorescence intensities. *The Journal of cell biology*. 209:163–180.

Cheezum, M.K., W.F. Walker and W.H. Guilford. 2001. Quantitative comparison of algorithms for tracking single fluorescent particles. *Biophysical journal.* 81:2378–2388.

Chuang, K.S., Chen, S and Hwang, I.M. 2001. Thresholding Histogram Equalization. *Journal of Digital Imaging.* 14:182–185.

Collins, T.J. 2007. ImageJ for microscopy. *BioTechniques.* 43:25–30.

Cordlières, F.P. 2005. Manual Tracking. Insitut Curie, Orsay (France) [accessed in 2015 June] Available in: http://rsbweb.nib.gov/ij/plugins/plugins/track/Manual%20Tracking%20plugin.pdf.

Cordlières, F.P., V. Petit, M. Kumasaka, O. Debeir, V. Letort, S.J. Gallagher and L. Larue. 2013. Automated cell tracking and analysis in phase-contrast videos (iTrack4U): development of Java software based on combined mean-shift processes. *PloS one.* 8:e81266.

Debeir, O., I. Adanja, R. Kiss and C. Decaestecker. 2008. Models of cancer cell migration and cellular imaging and analysis. *The Motile Actin System in Health and Disease:*123–156.

Gorelik, R. and A. Gautreau. 2014. Quantitative and unbiased analysis of directional persistence in cell migration. *Nature protocols.* 9:1931–1943.

Huth, J., M. Buchholz, J.M. Kraus, M. Schmucker, G. von Wichert, D. Krndija, T. Seufferlein, T.M. Gress and H.A. Kestler. 2010. Significantly improved precision of cell migration analysis in time-lapse video microscopy through use of a fully automated tracking system. *BMC cell biology.* 11:24.

Jaeger, S., Q. Song and S.S. Chen. 2009. DYNAMIK: a software environment for cell DYNAmics, Motility, and Information tracKing, with an application to Ras pathways. *Bioinformatics (Oxford, England).* 25:2383–2388.

Jain, P., R.A. Worthylake and S.K. Alahari. 2012. Quantitative analysis of random migration of cells using time-lapse video microscopy. *Journal of visualized experiments: JoVE:*e3585.

Kalaidzidis, Y. 2009. Multiple objects tracking in fluorescence microscopy. *Journal of mathematical biology.* 58:57–80.

Li, K., E.D. Miller, L.E. Weiss, P.G. Campbell and T. Kanade. 2006. Online tracking of migrating and proliferating cells imaged with phase-contrast microscopy. *In* Computer Vision and Pattern Recognition Workshop, 2006. CVPRW'06. Conference on. IEEE. 65–65.

Maeder, R.E. 2014. The Mathematica® Programmer. Elsevier Science.

Meijering, E. 2006. MTrackJ. Biomedical Imaging Group Rotterdam of the Erasmus University Medical Center Rotterdam, Netherlands. [accessed in 2015 July] imagescience.org/meijering/software/mtrackj/.

Meijering, E., O. Dzyubachyk and I. Smal. 2012. Methods for cell and particle tracking. *Methods in enzymology.* 504:183–200.

Meijering, E., I. Smal, O. Dzyubachyk and J.-C. Olivo-Marin. 2008. Time-lapse imaging. *Microscope Image Processing:*401–440.

Middleton, C.A. and J.A. Sharp. 1984. Cell Locomotion in Vitro: Techniques and Observations. University of California Press.

Miura, K. 2005. Tracking movement in cell biology. *Advances in biochemical engineering/biotechnology.* 95:267–295.

Möller, M., M. Burger, P. Dieterich and A. Schwab. 2014. A framework for automated cell tracking in phase contrast microscopic videos based on normal velocities. *Journal of Visual Communication and Image Representation.* 25:396–409.

Pawley, J. 2010. *Handbook of Biological Confocal Microscopy*. Springer US.

Peskin, A., D. Hoeppner and C. Stuelten. 2011. Segmentation and Cell Tracking of Breast Cancer Cells. *In* Advances in Visual Computing. vol. 6938. G. Bebis, R. Boyle, B. Parvin, D. Koracin, S. Wang, K. Kyungnam, B. Benes, K. Moreland, C. Borst, S. DiVerdi, C. Yi-Jen and J. Ming, editors. Springer Berlin Heidelberg. pp. 381–391.

Sacan, A., H. Ferhatosmanoglu and H. Coskun. 2008. CellTrack: an open-source software for cell tracking and motility analysis. *Bioinformatics (Oxford, England)*. 24:1647–1649.

Schindelin, J., I. Arganda-Carreras, E. Frise, V. Kaynig, M. Longair, T. Pietzsch, S. Preibisch, C. Rueden, S. Saalfeld, B. Schmid, J.Y. Tinevez, D.J. White, V. Hartenstein, K. Eliceiri, P. Tomancak and A. Cardona. 2012. Fiji: an open-source platform for biological-image analysis. *Nature methods*. 9:676–682.

Schneider, C.A., W.S. Rasband and K.W. Eliceiri. 2012. NIH Image to ImageJ: 25 years of image analysis. *Nature methods*. 9:671–675.

Xiong, Y. and P.A. Iglesias. 2010. Tools for analysing cell shape changes during chemotaxis. *Integrative biology: quantitative biosciences from nano to macro*. 2:561–567.

Zicha, D., G. Dunn and G. Jones. 1997. Analysing chemotaxis using the Dunn direct-viewing chamber. *Methods in molecular biology (Clifton, N.J.)*. 75:449–457.

8

Super-Resolution Data Analysis

Debora Keller[1], Nicolas Olivier[2], Thomas Pengo[3] and Graeme Ball[4]

[1] *Facility for Imaging by Light Microscopy, Imperial College London, UK*
[2] *Department of Physics and Astronomy, University of Sheffield, UK*
[3] *University of Minnesota Informatics Institute, University of Minnesota Twin Cities, USA*
[4] *Dundee Imaging Facility, School of Life Sciences, University of Dundee, UK*

8.1 INTRODUCTION TO SUPER-RESOLUTION MICROSCOPY

From 'Method of the Year' in 2008 to the Nobel Prize for Chemistry in 2014 super-resolution microscopy (SRM) has transitioned from a proof-of-principle method restricted to physicists to becoming available to biologists on a near-daily basis.

SRM has opened doors for biologists to study structures that were previously beyond the reach of conventional fluorescence microscopy because of the diffraction barrier. Briefly, the diffraction barrier relates to the fact that light cannot be focused into infinitely small spots because of diffraction. This limit is usually characterised by measuring the point-spread function (PSF) of a microscope, which corresponds to the image of a single point source after it has passed through the optical elements. The size of this PSF depends on two things: the emission wavelength, and the numerical aperture (NA) of the objective, which is an angular measure representing how tightly the light can be focused. Typically

Standard and Super-Resolution Bioimaging Data Analysis: A Primer, First Edition.
Edited by Ann Wheeler and Ricardo Henriques.
© 2018 John Wiley & Sons Ltd. Published 2018 by John Wiley & Sons Ltd.

(using 488 nm excitation), the resolution is limited to ≈ 250 nm in xy and ≈ 550 nm in the z dimension, thus objects that are smaller than this cannot be resolved.

There are several existing super-resolution methods, which try to bypass the diffraction limit, either by modulating the excitation in time or patterning it in space. The resolution achieved varies greatly between the different methods. For example, structured illumination microscopy (SIM) can achieve half the conventional resolution limit to reach 120 nm laterally and 300 nm axially, while stimulated emission depletion (STED) microscopy can resolve objects of 40–70 nm in the lateral and 40–150 nm in the axial dimension. Finally, localisation-based methods such as 3D-stochastic optical reconstruction microscopy (3D-STORM) currently yield the best resolution, with ≈ 10–30 nm and ≈ 10–70 nm in the lateral and axial dimensions respectively.

The principles, requirements, limitations and applications of those super-resolution techniques have been covered in numerous excellent reviews (Huang et al., 2010; Schermelleh et al., 2010; Galbraith and Galbraith, 2011; Sydor et al., 2015).

This chapter will provide an introduction to the analysis and quantification of super-resolution data with an emphasis on single molecule localisation microscopy methods (SMLM – encompassing PALM, (d) STORM and GSDIM) and structured illumination techniques. We will not discuss STED image analysis here because STED uses optical methods to 'slim down' the PSF used for raster scanning an image. The outputs can be analysed in a similar fashion to confocal microscopy so the interested reader should consult chapters in other publications, which discuss best practice with STED microscopy (Revelo and Rizzoli 2015).

We will discuss both sieving and quality control on the raw and super-resolved images before going into the different quantifiable features, mentioning necessary controls and potential pitfalls.

8.2 PROCESSING STRUCTURED ILLUMINATION MICROSCOPY DATA

Structured illumination microscopy (SIM) uses spatially structured illumination and image processing to achieve double the resolution implied by the classical limit of Abbe. Although the resolution improvement is limited relative to STED and SMLM, SIM has several important advantages: it is fast, sensitive and does not require special fluorophores or sample preparation. Acquiring multichannel 3D SIM images is reasonably straightforward, and live-cell imaging is feasible. This section describes the process of reconstructing raw SIM data to obtain a super-resolution image and crucially, assessing the quality of these results. Finally, some of the studies that have successfully used SIM data to answer important biological questions are discussed.

8.2.1 SIM Reconstruction Theory

The smallest distance between two objects that can be resolved by a diffraction-limited system such as a fluorescence microscope is approximately half the wavelength of the light emitted. SIM uses spatially structured illumination (Gustafsson, 2000). This leads to frequency mixing between the illumination pattern and spatial frequencies (i.e. structures) in the sample. The resulting beat patterns, or Moiré fringes (Figure 8.1a), are larger than the structures they depend on, and can therefore be captured by the imaging system. Figure 8.1b illustrates that, in order to achieve an isotropic sampling of an extended frequency range that

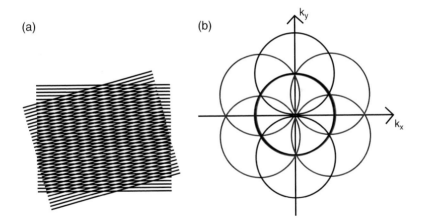

(a)　(b)

Figure 8.1 Structured Illumination increases resolution by extending the effective observable frequency range. Structured illumination generates frequencies that the imaging system can capture by frequency mixing between the striped (sinusoidal) illumination pattern and subresolution structures in the sample: these are analogous to the Moiré fringes (large vertical stripes) shown in (a). For the 2D case considered here (b), the illumination pattern contains three Fourier components. The central zero order component corresponds to frequencies observable in a standard wide-field image. Three illumination pattern phases are required to separate the three components shown in black, where the outer circles represent the region of the frequency space sampled by first-order components. The red circles represent two additional angles, which are required for isotropic sampling of all xy frequencies.

corresponds to a full higher resolution image, reconstruction of several overlapping subregions of the extended frequency range is necessary. The imaging system can only capture a limited range of frequencies, depending on its numerical aperture, but by varying the phase and angle of the illumination pattern, it is possible to effectively change the sub-region of the extended range of frequencies that are sampled in any one image. Since the higher frequencies have been displaced in order to be observable, they must be shifted back to the correct position in frequency space to produce the final super-resolution image. Although Figure 8.1 illustrates the 2D case only, the sample principle is used for 3D SIM, which relies on a 3D illumination pattern containing five orders (five phases are therefore required).

8.2.2 *Parameter Fitting and Corrections*

Three-dimensional structured illumination data can be combined through a generalised Wiener filter to obtain a reconstructed image (Gustafsson et al., 2008):

$$\hat{\tilde{S}}(k) = \frac{\sum\limits_{d,m} O_m^{*}(\mathbf{k} + m\mathbf{p}_d) \tilde{D}_{d,m}(\mathbf{k} + m\mathbf{p}_d)}{\sum\limits_{d',m'} \left| O_{m'}(\mathbf{k} + m'\mathbf{p}_{d'}) \right|^2 + w^2} A(k) \tag{8.1}$$

where S is the estimated true sample image, $\mathbf{k}$ are wave vectors, O are the optical transfer function (OTF) components, D are unshifted information components obtained by imaging the sample, $\mathbf{p}$ is the illumination pattern vector, m is order, d is pattern orientation, A is an apodisation function and w^2 is the Wiener filter parameter. The OTF must be measured using subresolution beads prior to SIM imaging, and its quality has a strong impact on the quality of the reconstruction. The pattern wave vector $\mathbf{p}$ (including starting phase and modulation amplitude) is generally fitted from each SIM dataset by cross-correlation between the different components where they overlap. The three repeats of the zero order component can also be used to correct for drift during the course of the experiment. The Wiener parameter can be used to boost or suppress high frequencies, and its optimal value depends on the amount of high frequency information available: where signal-to-noise and/or contrast are poor, enhancing high frequencies will simply generate noise artefacts. Recently, free open-source software toolboxes for reconstructing SIM data have begun to appear (Müller et al., 2016, Křížek et al., 2016) in addition to software offered by SIM instrument manufacturers.

8.2.3 SIM Quality Control

The reconstruction procedure described above is sensitive to a number of potential defects in the sample, the imaging system and the parameters chosen. A variety of artefacts can be generated by a poor reconstruction, leading to degraded image quality and perhaps even erroneous conclusions. It is therefore essential to assess the quality of the reconstruction, and be aware of the common problems and their remedies. Ball et al. (2015) describe an ImageJ plugin suite, SIMcheck (Figure 8.2a and Table 8.1), which can be used to assess the issues discussed below.

8.2.4 Checking System Calibration

High quality SIM data depends on a well-aligned system. Although this will not be discussed in detail here, it is worth noting that the structured illumination pattern must be properly focused (e.g. SIMcheck illumination pattern focus check using a bead lawn sample). Furthermore, all components must be present and of sufficient amplitude: in a 3D experiment, this means first-order and second-order Fourier components. To assess this, record raw SIM data for a sample that fills a large fraction of the field of view and examine its Fourier transform (in SIMcheck, this is presented as a projection of the Fourier amplitudes, Figure 8.2b). Finally, it is important that the illumination pattern phase has a stable offset and steps with the correct uniform step size. Again, this can be assessed using a Fourier transform of raw SIM data using, for example, a bead lawn sample. In this case, the phase of first-order spots can be used to monitor the phase as it is stepped in the raw data (Ball et al., 2015).

8.2.5 Checking Raw Data

When acquiring SIM data, there are a number of common pitfalls: spherical aberration, sample movement, photobleaching and insufficient contrast. Spherical aberration depends on sample refractive index (RI), depth and temperature; it should be minimised as much as possible to achieve good reconstructions. In the absence of a motorised correction collar, a workaround to minimise spherical aberration is to choose the optimal RI oil for the wavelength and depth of interest (generally by observing the z symmetry of a point source of light in the sample). Movement of a sample during a SIM acquisition can lead to stripe

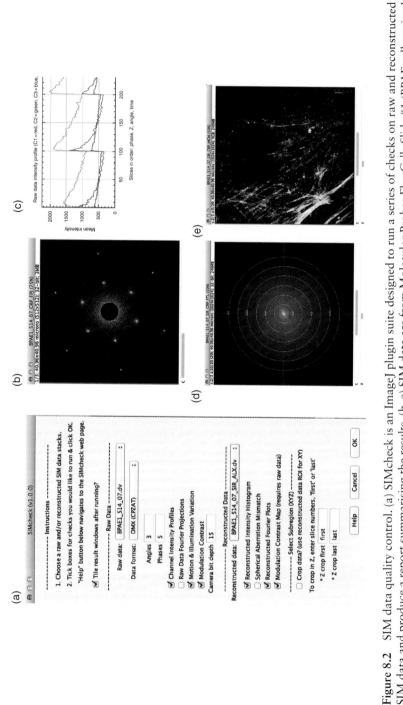

Figure 8.2 SIM data quality control. (a) SIMcheck is an ImageJ plugin suite designed to run a series of checks on raw and reconstructed SIM data and produce a report summarising the results. (b–e) SIM data are from Molecular Probes FluoCells Slide #1 (BPAE cells stained with DAPI, Alexa Fluor 488 Phalloidin and MitoTracker Red), acquired by Paul Appleton, on a GE OMX V4 Blaze (Dundee Imaging Facility): (b) shows a projection of Fourier transformed raw data channel 1 (DAPI) with the zero order spot blanked out, first- and second-order spots clearly visible, (c) shows average (slice) intensity profiles for the raw data as phase, z position and angle are incremented, (d) shows a 2D Fourier transform of the channel 2 (phalloidin) reconstruction, with 'resolution rings' in microns, (e) shows the channel 2 reconstruction colour-coded according to modulation contrast-to-noise (mostly orange-yellow, indicating >10, which is good).

Table 8.1 SIM data quality checks (SIMcheck).

Check	Type	Statistic/Comments
Illumination pattern focus	Calibration	Inspect pattern focus
Illumination phase steps	Calibration	Check phase step; check offset stability visually
Intensity profiles	Raw data	Total intensity variation
Motion/illumination Variation	Raw data	Differences between angles are coloured
Fourier projection	Calibration/Raw data	Check for first and second order spots
Modulation contrast	Raw data	Modulation contrast-to-noise ratio (MCNR) >3
Intensity histogram	Reconstructed	Min-to-max intensity ratio >3
Spherical aberration mismatch	Reconstructed	z minimum variation
Fourier transform	Reconstructed	Inspect for artefacts and to assess resolution

artefacts in the reconstructed image if it occurs on a timescale shorter than the time necessary to acquire all phases and angles for a given z section and its immediate neighbours. Photobleaching can lead to two problems (1) Severe bleaching that creates large intensity differences over a small Z window that have nothing to do with structures in the sample (i.e. as phase and angle is varied) can lead to reconstruction artefacts, (2) More commonly, the cumulative effects of photobleaching can lead to weak signals and insufficient signal-to-noise in the final images of a series. The Channel Intensity Profiles check of SIMcheck (Figure 8.2c) is designed to assess both. Finally, a critical requirement for successful reconstruction of SIM data is contrast; typically this limits SIM to the first 20 μm or so of a sample, depending on the density of structures. A useful metric to calculate is the modulation contrast, i.e. the amplitude of second- and/or first-order components of the structured illumination pattern relative to the zero-order component (visually this corresponds to distinct stripes in the raw data!) SIMcheck can be used to calculate modulation contrast relative to estimated noise, and furthermore convert this to a colour lookup table overlaid on the reconstructed image (Figure 8.2e). This can help to judge whether structures observed in the reconstructed image are supported by reliable raw data or are more likely to be noise artefacts.

8.2.6 *Checking Reconstructed Data*

The intensity histogram of a reconstructed SIM image can be a very simple indicator of reconstruction quality, provided that the result has not been clipped (some reconstruction software has an option to 'discard negative values' or similar). Where the shape of the image histogram is skewed towards positive/above-mode intensities, this indicates real intensity information has been recovered. If the histogram is very symmetrical about the zero/modal intensity value, this indicates a poor reconstruction and noise artefacts. A further assessment of the quality of the reconstruction can be made by examining Fourier amplitudes for the reconstructed data (see Figure 8.2d for SIMcheck output). A high-quality reconstruction will generally decay smoothly from the centre of the Fourier amplitude image towards the higher frequencies at the edges without a visible sharp edge or spots. Where spots can be seen in this Fourier transform, corresponding hexagonal arrays of artefacts are usually observed around real features in the reconstructed image. Where the Fourier amplitudes decay rapidly to a plateau, this is the limit of real high frequency information, beyond which only noise is present: in which case, this should be suppressed by a stronger Wiener filter during reconstruction, accepting a more blurred, lower-resolution final image free of noise artefacts.

8.2.7 SIM Data Analysis

Other than the reconstruction procedure and its quality control outlined above, analysis of SIM data tends to be very similar to the analysis of 'standard' fluorescence microscopy data. The main differences are that: (1) the higher resolution enables features to be resolved that could not be resolved previously, and (2) the higher resolution in all three dimensions creates reconstructed datasets that are eight times larger than a standard fluorescence microscopy image of the same volume (the raw data are typically around a factor of two larger again, i.e. 15 times larger). This order of magnitude increase in data is enough to create its own 'big data' challenge. Finally, of the three super-resolution techniques presented in this chapter, SIM is perhaps the most compatible with live-cell imaging: being a wide-field technique it is relatively fast and non-photodamaging compared to STED and SMLM.

The twofold resolution increase and improvement in contrast afforded by SIM can reveal new structural details and insights into biological processes. Initial SIM publications with biological data showed that SIM could resolve details such as the maize meiocyte synaptonemal complex (Gustafsson et al., 2008), mitochondria (Hirvonen et al., 2009) and the nuclear pore complex (Schermelleh et al., 2008) that could not be properly resolved by standard wide-field or confocal laser scanning microscopy. Note that, in contrast to SMLM and STED, standard fluorophores can be used to obtain multichannel SIM images combining three or four different labels with relative ease. This makes SIM a good technique for correlating the structural details of several different molecules (provided the resolution is sufficient, of course). For example, Lawo et al. (2012) used 3D SIM to uncover a layered organisation of centrosome components by imaging different pairs of molecules simultaneously.

The 2009 studies of Hirvonen et al. and Kner at al. (2009) showed the feasibility of live SIM imaging (Hirvonen et al. followed the arrangement of mitochondria in Cos cells labelled with MitoTracker over three three-minute intervals; Kner et al. imaged tubulin and kinesin dynamics in total internal reflection mode). Hardware developments soon enabled true live 3D SIM data to be acquired: Shao et al. reported in 2011 that they were able to follow microtubule dynamics in Drosophila S2 cells and mitochondrial network dynamics in HeLa cells with 120 nm lateral, 360 nm axial and 5 s temporal resolution. Since then, the achievable temporal resolution has continued to improve to the subsecond level with developments in CMOS camera and spatial light modulator (SLM) technologies.

Two examples of the application of live super-resolution SIM imaging to biological problems are the publications of Ashdown et al. (2014) and Lesterlin et al. (2014). Ashdown et al. followed filamentous actin dynamics in the T cell immunological synapse via total internal reflection mode (2D) SIM. They performed spatiotemporal image correlation spectroscopy (STICS) using SIM data to quantify actin flow on subresolution length scales in live cells, and found the flow to be retrograde and radially directed throughout the periphery of T-cells during synapse formation. Lesterlin et al. used 3D SIM in both fixed and live E.coli samples to follow the structure and dynamics of RecA, a DNA recombination protein, and chromosomal loci during double strand break repair. SIM data revealed an unexpected nucleation of RecA bundles at the site of double stranded breaks, followed by extension of these bundles along the long axis of the bacterial cell to mediate homology pairing between distant sister chromatids.

8.3 QUANTIFYING SINGLE MOLECULE LOCALISATION MICROSCOPY DATA

Single-molecule localisation fluorescence microscopy (SMLM) consists in essence of creating a map of probable locations for individual fluorophores. After acquiring a sequence of images at a high frame-rate where fluorophores are cycled through one or multiple bright and dark states, the sequence is processed to produce a list of fluorophore positions and a probability density map for their distribution in the sample. The process from acquiring the image sequence to generating a quantifiable map can be split into three main steps – pre-processing, localisation and rendering – that have to be done before analysing the super-resolved image and extracting quantifiable parameters. Figure 8.3 illustrates schematically the different steps from raw images to super-resolved image.

Pre-processing tries to mitigate noise and reduce background; localisation identifies bright spots and estimates the position of each emitter; rendering creates the density map, the super-resolved image, from the list of fluorophores that will be used for subsequent image analysis. The crucial step in any localisation workflow is the localisation step itself.

8.3.1 SMLMS Pre-Processing

Before the raw images are fed to a localisation algorithm, an optional pre-processing step is possible. Pre-processing such as background correction or smoothing should be applied with caution since it may introduce localisation artefacts. However, in some cases it can improve the images. In particular, large backgrounds can be efficiently subtracted to retrieve information from an extended structure along the z axis, and inhomogeneous camera gain can be corrected (Huang et al., 2013).

8.3.2 Localisation: Finding Molecule Positions

8.3.3 Fitting Molecules

The goal of this step is to identify the position of each emitter with sub-pixel precision. The sequence of images is in most cases analysed frame by frame. The classic approach is to identify regions of local maxima using well-established algorithms, extract a small window of pixels around those maxima and then fitting a model function to the pattern. The majority of software packages for 2D SMLM data use a Gaussian function as model to fit the patterns, which we will describe in a bit more

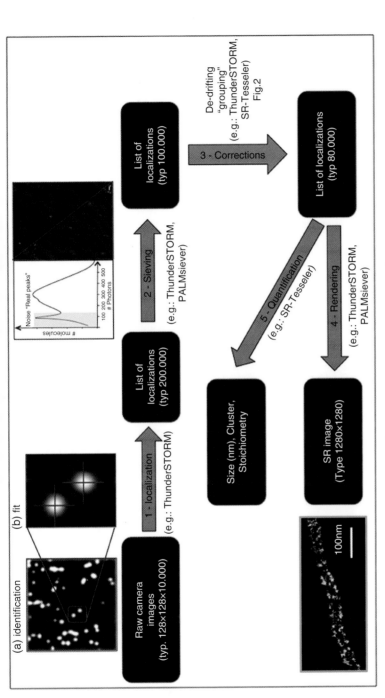

Figure 8.3 Sequence of steps from raw SMLM images to reconstructed super-resolved image. Examples of software as well as typical sample type are indicated for each step. Typically 10,000 raw images containing single molecule events are acquired, followed by (1) localisation of the centre of each molecule of each frame through a fitting process resulting in a list of localisations, (2) subsequent sieving of this list discards noise from 'real peaks' to generate a curated list of ~100,000 localisations, (3) additional corrections such as de-drifting or grouping of multiple emitters (see also Figure 8.4) are performed before (4) rendering of a super-resolved (SR) image or (5) quantification of specific features, e.g. structure size, cluster organisation or stoichiometry.

detail below. The principle of fitting for 3D SMLM data is similar, but the algorithms will use more complex model functions depending on how 3D was achieved on the specific instrument. For instance, commercial platforms such as Nikon or Zeiss use different means of PSF shaping to achieve 3D SMLM data (astigmatism or bi-lobed shape respectively) and as a consequence users can be restricted to using specific fitting algorithms implemented in their software. However, parameters important for localising 2D data also remain valid for 3D data (see below).

Any model-fitting algorithm iteratively searches the model parameter space and finds the optimal set of parameters according to a measurable criterion. The criterion is a cost function that quantifies the distance between the estimation and the data. This is often the sum of the squared error, but a more statistically effective way is to maximise the likelihood of the estimation (maximum likelihood estimation). Under the hypothesis of additive Gaussian noise, the two methods coincide. This only holds for large numbers of photons, so when detecting weak signals, the Poisson model or more complicated noise models (EMCCD) are better suited and yield a higher accuracy, but require more computational effort.

Alternatively, instead of performing a full fit – which, repeated hundred of thousands of times, can be time-consuming – a few algorithms perform fast approximations, such as calculating the centre of mass (e.g. QuickPALM), the centre of radial symmetry (e.g. RadialSymmetry) or triangulation (e.g. fluoroBancroft). In high signal-to-noise situations these approximations are in very good agreement with the more advanced estimations, with a significant increase in speed. Parallelisation, GPU computing and optimised code have also proven to be successful (e.g. GPUgaussMLE, RapidSTORM), enabling the more computationally intensive methods also to be used effectively.

Another family of algorithms avoids performing the fit completely, and iteratively estimates a higher resolution image directly, using deconvolution or sparse coding (FALCON).

The performance and advantages of the different localisation software for both high-density or sparse datasets has been evaluated thoroughly within the public 'Localisation Challenges' (http://bigwww.epfl.ch/smlm/) and recently published (Sage et al., 2015).

8.3.4 Problem of Multiple Emissions Per Molecule

Ideally, a single fluorophore would appear only in a single frame of the sequence. However, depending on the nature of the imaging experiment, it may be difficult to achieve this completely. The switching behaviour is a stochastic process that depends on the photophysics of each dye or fluorescent protein (FP) as well as the local environment, so on and off times vary from one molecule to another (Dempsey et al., 2011). This effect is especially important to take into consideration when performing counting or clustering analyses, as it can result in over-counting (see below).

To mitigate this issue, special post-processing techniques allow the combination of many different appearances of a single molecule into one, so that a single fluorophore is counted only once. This operation is often referred to as 'grouping' or 'blink correction' (Puchner et al., 2013; Durisic et al., 2014). It is analogous to tracking and clustering, except for a couple of important considerations. Grouping differs from tracking in that the emitter is assumed not to move; it can thus be considered to be a form of spatiotemporal clustering of the localisations, but the distance metric must take spatial and temporal dimensions into account separately.

The basic algorithm consists in looking into the near future of each localisation (within a few frames, Figure 8.4a) for possible additional appearances within a certain distance ('grouping radius' r_g of the order of the localisation precision; Figure 8.4b) and then grouping them into a single point obtained by a photon-weighted average of the position of all the elements of that group (Figure 8.4). The maximum time that a molecule is permitted to be off between appearances, and the maximum distance between appearances, are best estimated from control samples with single emitters (Annibale et al., 2011b; Puchner et al., 2013; Sengupta et al., 2011, 2013a).

8.3.5 *Sieving and Quality Control and Drift Correction*

Sieving refers to the process of filtering SMLM data in order to discard invalid localisations and just keeping the ones corresponding to set criteria. These invalid localisations arise from several sources discussed below.

In the analysis of sequences from multicolour experiments, additional care is needed at least in the following two aspects: spectral overlap and chromatic aberrations. The first is due to overlap between emission and excitation spectra of two fluorophores, which hinders the assignment of each point localisation to the appropriate fluorophore. The second causes slight distortions of the image through the objective in a wavelength-dependent manner, which hinders the ability to correctly identify the relative position of emitters from fluorophores with different emission

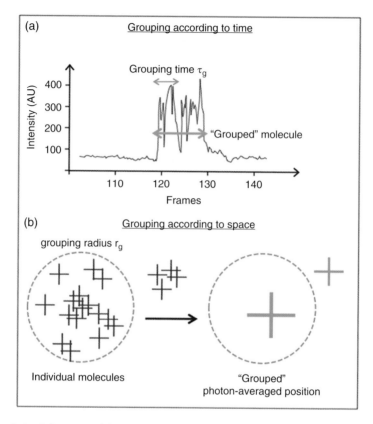

Figure 8.4   Schematic of the 'grouping' or 'blink-correction' procedure. (a) Temporal and (b) spatial intensity traces of a small image region showing multiple emission peaks over several frames, characteristic of a single molecule blinking behaviour. All the localisations within a given region of space limited by a grouping radius r_g (b), and time limited by a grouping time τ_g (a) are considered to originate from the same molecule and their positions are averaged into a new grouped photon-weighted average molecular localisation.

wavelengths. Both can and should be corrected in order to obtain meaningful results from the experiment.

Spectral overlap can be corrected effectively whenever the acquisition of two different wavelength windows (channels) is simultaneous. The different fluorophores will have distinct intensities in the recorded channels, which is then used to determine the most likely emitter. In fact, this method has been used to distinguish more than two fluorophores (Gunewardene et al., 2011).

Chromatic aberration is trickier to correct. High-end objectives are corrected for chromatic aberrations but only up to a certain resolution and in a wavelength-dependent manner. Correction of chromatic aberrations (or channel alignment) can be performed by first estimating the distortion and then applying the correction to the localisations. To estimate the distortion, the accepted practice is to use beads of a size below that of the diffraction limit of the microscope, coated with multiple fluorophores of different colours. The beads are set on a coverslip and imaged in the same conditions as the experiment (same mounting medium, same optical setup). The pattern is recorded in all channels of interest and multiple axial positions. The stack of images from one of the channels is taken as reference, and the deformation is calculated with registration algorithms from each other channel to the reference channel. There are many registration algorithms, and the most effective ones allow deformations from one channel to the next, which are non-rigid (not only translations or rotations) and therefore allow to account for more subtle deformations of the optics (Erdelyi et al., 2013).

Drift correction: Although the acquisition of LM images can be sped up to around 1 second by using sCMOS cameras and high power lasers (Huang et al., 2013), for most typical users the total acquisition time will be measured in minutes. At this timescale, drift becomes a very limiting factor to achieve high resolution. Luckily, sample drift is usually low-frequency and can therefore be efficiently corrected by drift correction.

Several groups have studied this aspect, and two main families of drift correction have been identified: fiducial-based and image-based. Most super-resolution software offer drift correction as an option, sometimes both with fiducial and image-based methods. We sum up the options offered by some of the existing open access software in Table 8.2.

Fiducial-based drift correction: Fiducial markers are non-blinking subdiffraction emitters that are spiked onto the sample and give the easiest way to characterise and correct drift. Examples include gold or silver colloids, as well as fluorescent beads. In order to de-drift the image, the x, y (z) positions of several colloids are recorded as a function of time, averaged and typically processed using a low-pass filter to remove the low-frequency drift without introducing any high-frequency artefacts. How much the data is filtered consists in a trade-off between the correction efficiency and the risk of introducing unwanted artefacts. A main limitation of this method is that fiducials are usually randomly distributed on the sample, so they can either overlap with the structure if the density is too high (though that can be corrected using a multipeak fitting algorithm e.g. FALCON, see Table 8.2 and Sage et al., 2015), or be absent from the field of view if the density is too low. Several protocols can be found online on sample preparation with fiducials.

Table 8.2 Examples of SMLM analysis software and its features. This non-exhaustive list provides an overview of existing software and the included characteristics. Note: for 3B and FALCON, some grouping is performed by default by the software during the peak localisation step ('implicit') but grouping cannot be done separately.

Name/ Feature	3B	FALCON	rain STORM	rapid STORM	Thunder STORM	PeakFit	PALM siever	SR-Tesseler	Bayesian Clustering	micro Manager	Quick PALM
Localization	2D, HD	3D astigm, HD	3D astigm	3D astigm & biplane	3D astigm, HD	2D	No	No	No	2D	3D astigm.
Drift correction	No	No	Fiducial	Fiducial	Fiducial & CC	Fiducial	Fiducial & CC	No	No	Fiducial	Fiducial
Sieving	No	No	partial	No	Yes	Yes	Yes	No	No	partial	No
Grouping	implicit	implicit	No	No	Yes	Yes	Yes	Yes	No	Yes	No
Analysis	No	No	SPT, resolution estimate	No	colocalization, simulation	PC-PALM, blink correction, tracing	DBSCAN tracing, custom code	Clustering, Voronoi segmentation	Clustering	SPT	
Platform	Unix, ImageJ	MATLAB	MATLAB	Win/Unix	ImageJ	ImageJ	MATLAB	Win	R	ImageJ	ImageJ
Standalone?	exec and plugin	code	code	exec	plugin	plugin	exec	exec	code	plugin	plugin

Abbreviations: HD = high density; 3D astigm. = 3D astigmatism PSF shaping; CC = cross-correlation; SPT = single-particle tracking; code = source code available; exec = executable; plugin = downloadable ImageJ/Fiji plugin.

Image-based drift correction: Image-based drift correction is a more indirect method that relies on the data itself as a reference. Several approaches have been proposed, but the basic idea is that images are reconstructed using successive subsets of the data, and the images are cross-correlated to provide an average displacement between subsets. The displacement is then calculated for several time points, and often low-pass filtered as with fiducial-based methods. A typical number of frames used for partial reconstruction is 500 frames, but the number has to be adapted to the density of emitters and to the structure.

Distinguishing noise and real peaks: Once the localisation software is done localising all the peaks there comes the most difficult step of the SMLM imaging process: distinguishing 'real' localisations from noise. This part requires some *a priori* knowledge of what is expected, and will be different for each experiment. It is typically done by considering the distribution of the fitted parameters. We will consider here a simplified model of 2D SMLM, where there is no background and the output of the fitting algorithm only contains three variables: the number of photons (whether in photons or in arbitrary units), the sigma of the Gaussian fit (in nm), and an error term (in arbitrary units).

1. The sigma of the Gaussian fit of a molecule (also referred to as width of the Gaussian or PSF width) is the easiest parameter to use to remove unwanted peaks. Within a field of view, the sigma or PSF width should have a minimum value, which is set by the diffraction limit, and out-of-focus molecules will have a larger width. All the molecules that appear significantly smaller than the diffraction limit can therefore be discarded as noise, as well as molecules with a PSF width much larger than expected. After sieving a 2D image, we therefore expect a narrow asymmetric distribution of sigma values with most values around the diffraction limit, and a broader tail towards large values due to out-of-focus molecules.

2. The number of photons emitted by a single molecule follows a Poisson distribution. If the localisation software is set up to be very sensitive, it will also detect localisations with a very low number of photons. Since the localisation precision depends on the number of photons (on the square root of the number of photons), molecules with low photon counts can be discarded as their position would be imprecise. Molecules with a high number of photons will be localised with higher accuracy and therefore higher confidence. Typical threshold values (for a calibrated camera which provides photon counts) go from 100 to 250 for fluorescent proteins (Puchner et al., 2013) to

several thousands of photons for bright dyes (Olivier et al., 2013; Vaughan et al., 2012). However, discarding localisations with low photon counts will also decrease the number of molecules defining the structure of interest and therefore the density of points in the final image. The trade-off between localisation density and localisation precision is discussed in Section 8.3.6 (Betzig et al., 2006).

3. Most software outputs at least some form of error term for each localisation, corresponding to the imprecision of the fit (or measure of confidence of the fit). The definitions of this error term can vary quite widely between the different software packages. Nevertheless, it remains a useful filtering criterion, because it is not always correlated to the other two parameters: a large error can, for example, indicate a bright non-Gaussian spot that has roughly the right size, but the wrong shape (for example, due to several overlapping molecules). A threshold to remove large values, i.e. localisations with high uncertainties, is therefore typically used.

Some software outputs a lot more fitting parameters (e.g. ellipticity of the Gaussian fit, background variation) that can be used the same way. The previous paragraph is deliberately qualitative because good sieving requires *a priori* information about the expected distributions. This information can be gained, for example, by analysing a known structure (microtubules being popular) where the sieving can be correlated to the super-resolution image to estimate the expected distributions; or alternatively, by looking at areas devoid of structure, representing the 'noise', to get an idea of the parameters to filter out (e.g. outside the cell, non-specific signal).

8.3.6 *How Far Can I Trust the SMLM Data?*

Because the 'images' are so different from typical fluorescence microscopy methods, new methods need to be developed to understand how far the data can be trusted. In classical fluorescence microscopy methods, the resolution can be described by the PSF of the instrument, and any feature measured by the method that is larger than the resolution can be trusted. Currently, a major debate regarding SMLM is how 'resolution' should be defined to report data. SMLM data cannot be fully represented *a priori* by a single resolution value. The quality of an image depends on two things: (1) localisation precision, which is a distribution representing the uncertainty of each localisation, and that can be computed during the localisation process (Sage et al., 2015), (2) the density of molecules – also a distribution and usually a very broad one – which reflects both the nature of the sample observed as well as the stochasticity of the imaging process. The higher the density, the more accurately can local features be observed. As mentioned in the original PALM paper (Betzig et al., 2006) – these two parameters are not independent: the localisation precision can be increased almost arbitrarily by removing the low values from the distribution, but this means that the density of molecules will be reduced. This fundamental trade-off is the main driver for developing quantitative methods in SMLM analysis to maximise the information content of an image.

Nyquist and other resolution measures: One way to address the question of resolution has been to rely on tools used in other types of microscopy. Two main approaches have been developed to tackle the problem: local approaches and global approaches. In the global approach, the whole image is used to infer an average resolution, while the local approach emphasises the variability and is described in more details in the next paragraph. Two main forms of global approach have typically been used: density-based Nyquist criteria and cross-correlation methods.

Nyquist sampling theorem: This is used in conventional fluorescence microscopy to set the pixel size. In super-resolution microscopy, it can be used to try to define a density threshold over which the localisation precision could be used as a value for resolution (Betzig et al., 2006). Unfortunately, this approach does not really work because of the way intensity is encoded in the super-resolution image, and theoretical considerations have proved that, in the general case, the density would have to be impossibly high to fulfil the Nyquist criterion (Fitzgerald et al., 2012), and more generally that density is almost always the limiting factor in localisation microscopy.

Cross-correlation methods adapted from EM microscopy: Fourier ring correlation (FRC): This method relies on splitting the localised molecules into two sets of the same size randomly, creating two images and cross-correlating them. The more the images are correlated, the more trustworthy they are. This approach is quite robust and very convenient for comparing different algorithms or different dyes or different buffers in similar samples. However, it does not take into account the local changes in information content and cannot be used to predict whether a local feature is real or not (Pengo et al., 2015b).

Resolution as a confidence measure: Another approach to the resolution problem is to realise that the resolution of an image is in fact a wide distribution of values and that the average does not convey a lot of useful information. This means forgetting about the whole image – and therefore giving up on the possibility of comparing between different methods, and instead focussing on how much a given local measurement on the image can be trusted (Pengo et al., 2015b). Indeed, in most cases, this is why people rely on the resolution. The main advantage of this approach is that the density is much more clearly defined at the local level. The implementation of this method is somewhat similar to FRC, and relies on repeated measurement over repeated subsets of localised molecules. However, instead of looking at a general parameter such as the Fourier spectrum, a particular measurement is repeated, and its truthfulness assessed by a bootstrapped measurement confidence (BMC) defined as the standard deviation of the bootstrapped resampling distribution on the measurements. The main advantage of this method is that it is sensitive to changes in the local density (or localisation precision) and thus can give an idea of areas or structures in an image that are more 'trustworthy' than others. This output will be useful when performing quantitative measurements on SMLM data, as it could be used to only select features with high confidence or high resolution.

8.4 RECONSTRUCTION SUMMARY

As we have seen, several methods exist that give complementary information on resolution: if the aim is to compare the average quality of different methods or protocols, global methods are well adapted and provide an average resolution, but if the aim is to obtain a quantitative threshold over which the information is trustworthy, then local approaches are more relevant. It is also important to note that although density is very important for the quality of super-resolution images, methods that implicitly assume continuous structures will always overestimate the number of molecules needed to reconstruct a structure accurately. Since protein distributions are intrinsically discrete, methods that use this property will help extract all the possible information from the LM datasets.

8.5 IMAGE ANALYSIS ON LOCALISATION DATA

The advantage of SMLM image analysis is that it can provide information on your structure of interest that was previously unattainable, such as subresolution spatial organisation or heterogeneous non-random distributions ('clusters') or even the numbers and stoichiometry of molecules within a complex or macro-molecular structure. However, as for any method, care needs to be taken to avoid artefacts that can arise from the method (reviewed in Shivanandan et al., 2014; Durisic et al., 2014), or the sample itself (Whelan and Bell, 2015).

8.5.1 *Cluster Analysis*

The first and most important step in cluster analysis is to carefully group the localised molecules beforehand to avoid artificial clustering due to multiple emissions of a single molecule (discussed above). Multiple emissions will occur in STORM samples (where single dyes are reactivated multiple times) but also in PALM experiments, as has been shown for several switchable FPs (mEos, mEos2, PA-GFP; Annibale et al., 2011a; Puchner et al., 2013). Once the localisations have been grouped, de-drifted and sieved, several algorithms are available to analyse clustering. These are summarised here and discussed at greater length in Chapter 6.

The simplest algorithms to identify clusters are density-based such as Ripley's K (Owen et al., 2010) or DBSCAN (density-based spatial clustering of applications with noise; Endesfelder et al., 2013). Both algorithms compute the probability of points to be clustered within a certain area, as opposed to a random distribution. When repeated over each localisation of an image, they result in a distribution function or histogram that peak at radii where points are clustered above random, enabling the average size of those clusters to be determined. The analysis with DBSCAN, for instance, can identify clusters of varied sizes (Endesfelder et al., 2013) and this is incorporated into tools such as PALMsiever which allows the user to go from filtering of his data, grouping and cluster analysis to rendering of the nanoscale clusters (Pengo et al., 2015a).

More advanced analysis tools such as pair-correlation analysis with PALM data (PC-PALM; Sengupta et al., 2011, 2013b) uses, as its name indicates, a statistical pair-correlation function to identify clusters, their size, density and the number of proteins from PALM data (see Section 8.5.2). To correct for multiple appearances of a single FP due to 'blinking', and to avoid over-counting molecules within clusters, it is important to have calibration samples such as isolated FPs or non-clustered biological controls with the same FP. Using these controls and the localisation uncertainty of the fitting, PC-PALM can help distinguish artificial or random cluster of proteins, which will be autocorrelated and the pair-correlation function will be ~1, from real biologically relevant clusters, where the autocorrelation function will be greater than 1 for a given cluster radius. PC-PALM analysis has the advantage that it can be applied to two-colour PALM images and assess the degree of co-organisation or co-localisation of two different proteins (Sengupta et al., 2011).

Recently, the SR-Tesseler tool was published for both the segmentation and analysis of SMLM images (Levet et al., 2015). It is based on Voronoï diagram reconstruction, whereby individual polygons are centred around each localised peak of the image, resulting in a higher density of smaller polygons in clustered areas facilitating both the segmentation of the cluster and quantification of related features (size, shape, density, number of molecules).

8.5.2 Stoichiometry and Counting

Determining the stoichiometry and number of specific proteins within complexes (e.g. receptors) or macro-molecular assemblies (e.g. organelles) is key to furthering the understanding of biological processes. Traditional methods for estimating protein numbers include quantitative mass spectrometry (little information on spatial organisation), brightness analysis (limited to conventional resolution) or stepwise photobleaching (limited to low numbers) (Wu, 2005; Durisic et al., 2012).

The advent of PALM and engineering of irreversible photo-switchable or photoconvertible proteins (e.g. mEos, PA-GFP) has enabled scientists to get both a super-resolved structure and the corresponding protein stoichiometry. As mentioned earlier, it is important to bear in mind existing caveats to avoid under- or over-counting, as well as in interpreting the biological significance (overexpression function). For instance, it is estimated that 40–60% of tagged proteins are either not yet fluorescent or are missed due to imaging conditions, leading to under-counting (Durisic et al., 2014; Puchner et al., 2013). Similarly, labelling efficiency (Lau et al., 2012) or multiple-blinking (Annibale et al., 2011b) can lead to over-counting. It is thus crucial to incorporate appropriate experimental controls when counting molecules using super-resolution microscopy. In their elegant study, Puchner et al. used multiple repeats of membrane-localised mEos2 to calibrate their system for multiple blinking and to correct for undetected molecules and were able to count up to hundreds of molecules within their organelle of interest.

Counting in STORM is complicated by the fact that the photophysics of dyes is less well understood, with dyes cycling multiple times, and it is challenging to obtain a 1:1 degree of labelling (1 dye/primary antibody). However, using labelled RNA transcripts and normalising the blinking frequencies to a singly labelled probe allows the determination by STORM of the stoichiometry of Xist molecules in mouse nuclei (Sunwoo et al., 2015).

Clustering algorithms mentioned above, such as DBSCAN, PC-PALM or SR-Tesseler, can be used to extract counting histograms from the corrected data and to determine the stoichiometry of the clusters.

Taken together, clustering and stoichiometry analyses using super-resolved localisation data open up the possibilities for biologists to further their understanding of processes of interest, but extra care must be taken to include appropriate controls and avoid artefacts. Combining gene editing techniques to the development of new probes with higher photon yields, faster maturation time and diminished dimerisation (Wang et al., 2014) will further enhance the possibilities for SMLM in biology.

8.5.3 *Fitting and Particle Averaging*

The idea of fitting can be applied not only at the pattern level but also at the object level. If the structure of interest is, for example, a ring, then a ring can be fitted to the set of localisations deemed to be part of the same ring, thus allowing an estimate of the radius of the ring or the thickness of the rim (Sydor et al., 2015). Another analysis technique is that of particle averaging, taken from EM image analysis: if many copies of the same structure are imaged, the idea of particle averaging is to align all particles and combine the localisations to form an 'average' image.

Having the actual molecule positions allows the researcher to perform the analysis on the emitter positions rather than on the rendered images, eliminating a step of the process and potentially a source of bias.

8.5.4 *Tracing*

One elaboration of the concept of fitting is that of tracing. Whenever an elongated structure such as a microtubule is analysed, it is often of interest to analyse the cross-section of the microtubule. Instead of using small sections of the microtubules, limiting the signal-to-noise level of the profile, the estimation can be performed by averaging the profile along an extended section of the microtubule.

8.6 SUMMARY AND AVAILABLE TOOLS

Several tools exist for each step of SMLM image analysis, from fitting individual peaks to rendering and quantifying features, and as the field progresses, new software is being developed to address new questions. Table 8.2 provides a non-exhaustive list of existing SMLM software and its different available features.

8.7 REFERENCES

Annibale, P., S. Vanni, M. Scarselli, U. Rothlisberger and A. Radenovic. 2011a. Identification of clustering artifacts in photoactivated localization microscopy. *Nat Methods.* 8:527–528. doi:10.1038/nmeth.1627.

Annibale, P., S. Vanni, M. Scarselli, U. Rothlisberger and A. Radenovic. 2011b. Quantitative Photo Activated Localization Microscopy: Unraveling the Effects of Photoblinking. *PLoS One.* 6:e22678. doi:10.1371/journal.pone.0022678.

Ashdown, G.W., A. Cope, P.W. Wiseman and D.M. Owen 2014. Molecular flow quantified beyond the diffraction limit by spatiotemporal image correlation of structured illumination microscopy data. *Biophys J* 107(9): L21–23.

Ball, G., J. Demmerle, R. Kaufmann, I. Davis, I.M. Dobbie and L. Schermelleh 2015. SIMcheck: a Toolbox for Successful Super-resolution Structured Illumination Microscopy. *Sci Rep* 5: 15915.

Betzig, E., G.H. Patterson, R. Sougrat, O.W. Lindwasser, S. Olenych, J.S. Bonifacino, M.W. Davidson, J. Lippincott-Schwartz and H.F. Hess. 2006. Imaging intracellular fluorescent proteins at nanometer resolution. *Science.* 313:1642–1645. doi:10.1126/science.1127344.

Dempsey, G.T., J.C. Vaughan, K.H. Chen, M. Bates and X. Zhuang. 2011. Evaluation of fluorophores for optimal performance in localization-based super-resolution imaging. *Nat Methods.* 8:1027–1036. doi:10.1038/nmeth.1768.

Durisic, N., L.L. Cuervo and M. Lakadamyali. 2014. Quantitative super-resolution microscopy: pitfalls and strategies for image analysis. *Curr Opin Chem Biol.* 20:22–8. doi:10.1016/j.cbpa.2014.04.005.

Durisic, N., A.G. Godin, C.M. Wever, C.D. Heyes, M. Lakadamyali and J.A. Dent. 2012. Stoichiometry of the human glycine receptor revealed by direct subunit counting. *J Neurosci.* 32:12915–20. doi:10.1523/JNEUROSCI.2050-12.2012.

Endesfelder, U., K. Finan, S.J. Holden, P.R. Cook, A.N. Kapanidis and M. Heilemann. 2013. Multiscale spatial organization of RNA polymerase in Escherichia coli. *Biophys J.* 105:172–81. doi:10.1016/j.bpj.2013.05.048.

Erdelyi, M., E. Rees, D. Metcalf, G.S.K. Schierle, L. Dudas, J. Sinko, A.E. Knight and C.F. Kaminski. 2013. Correcting chromatic offset in multicolor super-resolution localization microscopy. *Opt Express.* 21:10978–88. doi:10.1364/OE.21.010978.

Fitzgerald, J.E., J. Lu and M.J. Schnitzer. 2012. Estimation theoretic measure of resolution for stochastic localization microscopy. *Phys Rev Lett.* 109:048102. doi:10.1103/PhysRevLett.109.048102.

Galbraith, C.G. and J.A. Galbraith. 2011. Super-resolution microscopy at a glance. *J Cell Sci.* 124:1607–11. doi:10.1242/jcs.080085.

Gunewardene, M.S., F.V Subach, T.J. Gould, G.P. Penoncello, M.V. Gudheti, V.V Verkhusha and S.T. Hess. 2011. Superresolution imaging of multiple fluorescent proteins with highly overlapping emission spectra in living cells. *Biophys J.* 101:1522–8. doi:10.1016/j.bpj.2011.07.049.

Gustafsson, M.G. 2000. Surpassing the lateral resolution limit by a factor of two using structured illumination microscopy. *J Microsc* 198(Pt 2): 82–87.

Gustafsson, M.G., L. Shao, P.M. Carlton, C.J. Wang, I.N. Golubovskaya, W.Z. Cande, D.A. Agard and J.W. Sedat 2008. Three-dimensional resolution doubling in wide-field fluorescence microscopy by structured illumination. *Biophys J* 94(12): 4957–4970.

Hirvonen, L.M., K. Wicker, O. Mandula and R. Heintzmann 2009. Structured illumination microscopy of a living cell. *Eur Biophys J* 38(6): 807–812.

Huang, B., H. Babcock and X. Zhuang. 2010. Breaking the Diffraction Barrier: Super-Resolution Imaging of Cells. *Cell.* 143:1047–1058. doi:10.1016/j.cell.2010.12.002.

Huang, F., T.M.P. Hartwich, F.E. Rivera-Molina, Y. Lin, W.C. Duim, J.J. Long, P.D. Uchil, J.R. Myers, M.A. Baird, W. Mothes, M.W. Davidson, D. Toomre and J. Bewersdorf. 2013. Video-rate nanoscopy using sCMOS camera-specific single-molecule localization algorithms. *Nat Methods.* 10:653–8. doi:10.1038/nmeth.2488.

Kner, P., B.B. Chhun, E.R. Griffis, L. Winoto and M.G. Gustafsson 2009. Super-resolution video microscopy of live cells by structured illumination. *Nat Methods* 6(5): 339–342.

Křížek, P., T. Lukeš, M. Ovesný, K. Fliegel and G.M. Hagen 2016. SIMToolbox: a MATLAB toolbox for structured illumination fluorescence microscopy. *Bioinformatics* 32(2): 318–320.

Lau, L., Y.L. Lee, S.J. Sahl, T. Stearns and W.E. Moerner. 2012. STED microscopy with optimized labeling density reveals 9-fold arrangement of a centriole protein. *Biophys J.* 102:2926–2935. doi:10.1016/j.bpj.2012.05.015.

Lawo, S., M. Hasegan, G.D. Gupta and L. Pelletier 2012. Subdiffraction imaging of centrosomes reveals higher-order organizational features of pericentriolar material. *Nat Cell Biol* 14(11): 1148–1158.

Lesterlin, C., G. Ball, L. Schermelleh and D.J. Sherratt 2014. RecA bundles mediate homology pairing between distant sisters during DNA break repair. *Nature* 506(7487): 249–253.

Levet, F., E. Hosy, A. Kechkar, C. Butler, A. Beghin, D. Choquet and J.-B. Sibarita. 2015. SR-Tesseler: a method to segment and quantify localization-based super-resolution microscopy data. *Nat Methods.* 12:1065–1071. doi:10.1038/nmeth.3579.

Müller, M., V. Mönkemöller, S. Hennig, W. Hübner, T. Huser 2016. Open-source image reconstruction of super-resolution structured illumination microscopy data in ImageJ. *Nature Comms.* 7:10980 doi:10.1038/ncomms10980.

Olivier, N., D. Keller, P. Gönczy and S. Manley. 2013. Resolution Doubling in 3D-STORM Imaging through Improved Buffers. *PLoS One.* 8.

Owen, D.M., C. Rentero, J. Rossy, A. Magenau, D. Williamson, M. Rodriguez and K. Gaus. 2010. PALM imaging and cluster analysis of protein heterogeneity at the cell surface. *J Biophotonics.* 3:446–54. doi:10.1002/jbio.200900089.

Pengo, T., S.J. Holden and S. Manley. 2015a. PALMsiever: a tool to turn raw data into results for single-molecule localization microscopy. *Bioinformatics.* 31:797–8. doi:10.1093/bioinformatics/btu720.

Pengo, T., N. Olivier and S. Manley. 2015b. Away from resolution, assessing the information content of super-resolution images.

Puchner, E.M., J.M. Walter, R. Kasper, B. Huang and W.A. Lim. 2013. Counting molecules in single organelles with superresolution microscopy allows tracking of the endosome maturation trajectory. *Proc Natl Acad Sci.* 110:16015–20. doi:10.1073/pnas.1309676110.

Revelo N.H, and S.O. Rizzoli 2015. Application of STED Microscopy to Cell Biology Questions Methods Mol Biol. 1251:213–30. doi: 10.1007/978-1-4939-2080-8_12.

Sage, D., H. Kirshner, T. Pengo, N. Stuurman, J. Min, S. Manley and M. Unser. 2015. Quantitative evaluation of software packages for single-molecule localization microscopy. *Nat Methods.* 12:717–724. doi:10.1038/nmeth.3442.

Schermelleh, L., P.M. Carlton, S. Haase, L. Shao, L. Winoto, P. Kner, B. Burke, M.C. Cardoso, D.A. Agard, M.G. Gustafsson, H. Leonhardt and J.W. Sedat 2008. Subdiffraction multicolor imaging of the nuclear periphery with 3D structured illumination microscopy. *Science* 320(5881): 1332–1336.

Schermelleh, L., R. Heintzmann and H. Leonhardt. 2010. A guide to super-resolution fluorescence microscopy. *J Cell Biol.* 190:165–75. doi:10.1083/jcb.201002018.

Shao, L., P. Kner, E.H. Rego and M.G. Gustafsson 2011. Super-resolution 3D microscopy of live whole cells using structured illumination. *Nat Methods* 8(12): 1044–1046.

Sengupta, P., T. Jovanovic-Talisman, D. Skoko, M. Renz, S.L. Veatch and J. Lippincott-Schwartz. 2011. Probing protein heterogeneity in the plasma membrane using PALM and pair correlation analysis. *Nat Methods.* 8:969–75. doi:10.1038/nmeth.1704.

Shivanandan, A., H. Deschout, M. Scarselli and A. Radenovic. 2014. Challenges in quantitative single molecule localization microscopy. *FEBS Lett.* doi:10.1016/j.febslet.2014.06.014.

Sunwoo, H., J.Y. Wu and J.T. Lee. 2015. The Xist RNA-PRC2 complex at 20-nm resolution reveals a low Xist stoichiometry and suggests a hit-and-run mechanism in mouse cells. *Proc Natl Acad Sci.* 1503690112. doi:10.1073/pnas.1503690112.

Sydor, A.M., K.J. Czymmek, E.M. Puchner and V. Mennella. 2015. Super-Resolution Microscopy: From Single Molecules to Supramolecular Assemblies. *Trends Cell Biol.* 25:730–748. doi:10.1016/j.tcb.2015.10.004.

Vaughan, J.C., S. Jia and X. Zhuang. 2012. Ultrabright photoactivatable fluorophores created by reductive caging. *Nat Methods.* 9:1–7. doi:10.1038/nmeth.2214.

Wang, S., J.R. Moffitt, G.T. Dempsey, X.S. Xie and X. Zhuang. 2014. Characterization and development of photoactivatable fluorescent proteins for single-molecule-based superresolution imaging. *Proc Natl Acad Sci.* 111:8452–7. doi:10.1073/pnas.1406593111.

Whelan, D.R. and T.D.M. Bell. 2015. Image artifacts in single molecule localization microscopy: why optimization of sample preparation protocols matters. *Sci Rep.* 5:7924. doi:10.1038/srep07924.

Wu, J.-Q. 2005. Counting Cytokinesis Proteins Globally and Locally in Fission Yeast. *Science* (80-). 310:310–314. doi:10.1126/science.1113230.

9

Big Data and Bio-Image Informatics: A Review of Software Technologies Available for Quantifying Large Datasets in Light-Microscopy

Ahmed Fetit*

Advanced Imaging Resource, MRC-IGMM, University of Edinburgh, UK
*Currently at: School of Science and Engineering, University of Dundee, UK

9.1 INTRODUCTION

The past two decades have seen unprecedented advances in microscopy acquisition technologies, empowering life scientists to monitor phenomena with extraordinary levels of resolution, richness and complexity [1]. Recording and acquisition of multiple sources of information (channels) has become routine practice in fluorescent imaging studies. With increasing complexity of the biological questions of interest, z-stacks of 2D images are easily incorporated into imaging workflows, across multiple time points, resulting in highly dimensional gigabytes of data in a typical imaging investigation [2]. Additionally, handling associated image metadata and annotations introduces extra layers of valuable information for understanding and quantifying imaging data.

Standard and Super-Resolution Bioimaging Data Analysis: A Primer, First Edition.
Edited by Ann Wheeler and Ricardo Henriques.

Such advancements in hardware and acquisition technologies should have a profound impact on a wide range of research areas, including human and animal genetics, genetic psychiatry, cancer research and digital pathology. Nevertheless, a number of barriers continue to impede momentum. Many life scientists in universities and research institutes remain uncertain about how to best orchestrate and analyse such increasingly rich datasets, and are often over- whelmed with the variety of options that could be utilised for their needs. Additionally, while a number of available software solutions do not particularly require deep technical familiarity of computer science principles, the technical terminology that is often used in documenting them presents additional challenges to users who do not come from computing intensive backgrounds. Finally, it remains noteworthy that a large number of currently available technologies require integration with sophisticated IT infrastructure (e.g. use of high performance computing clusters). While, in principle, setting up such technologies is conceptually straightforward for system administrators and IT sup- port professionals, the lack of a 'common language' that can be used between technical support personnel and biology researchers intro- duces additional layers of complexity.

These challenges are particularly apparent in the example of high- content screening, where information is gathered in high throughputs in order to formalise visual imaging characteristics that can be used to phenotype samples of interest. Use of routine 'manual' analysis pipelines, storage on local workstations and conventional exploration of results impose usability limitations with such complex workflows. This scenario is a practical example of what is referred to as a *big data* challenge – the focus of this chapter.

9.2 WHAT IS BIG DATA ANYWAY?

Before diving into details of big data in microscopy, it is necessary to take a step back and appreciate that the phenomenon is not only bioim- aging or healthcare centric, but is rather migrating to fundamentally influence all aspects of our society. While it would be helpful to provide a rigorous definition of what big data is, such characterisation does not presently exist [3]. Initially, the term was coined in the 2000s to express the observation that the volume of information available for processing had grown to such large extents that the quantities being examined no longer fitted the memory specifications that conventional computers used for processing. This data comes from everywhere: social media content, cell phone GPS signals, electronic health records and financial information to name but a few.

It is rather difficult to pinpoint an exact reason for the observed tremendous growth in data available to us, but as Mayer-Schonberger and Cukier [3] argue, the ability to record information is one of the lines of distinction between primitive and advanced societies. It is perhaps our way of quantifying the world around us. It can be observed when studying the world's earliest civilisations, where for example, ancient Egyptians maintained records and measurements of everyday life, from recording history to keeping track of business transactions. 'Quantification' facilitated simulation and hence planning for all aspects of life, from predicting floods to guessing how fruitful the next year's harvest would be. Over the centuries, the tools used to record data, and the type of data itself, had dramatically changed. However, with the rise of mathematics and the spreading of modern numerals to Europe, a new dimension to data escalated: data analysis. Back to the modern day, robust data analysis is now the ultimate goal of investment bankers trying to quantify the financial market, astrophysicists trying to understand the galaxy – and more relevant to this chapter – life scientists trying to phenotype biological samples using a microscopy system.

A concept that is often associated with the big data phenomenon is what technology professionals refer to as Moore's law. It is an observation made in 1965 by Intel's co-founder, Gordon Moore, and not actually a physical law. Moore noted that computing would dramatically increase in power while relative cost would diminish, and both at exponential rates. While the concept was initially coined in the form of a forecast, it later became widely accepted as a gold-standard by the semiconductor industry. Consequently, manufacturers placed their focus on meeting the processing power predicted by Moore's law because it was generally assumed that the competition would do the same. Such advancements in computer storage and processing capabilities which are now easily available to universities and research institutions, together with progressions in microscopy acquisition technologies, have paved the way for big data in bioimaging.

Despite the lack of a structured definition of what big data is, and the variety of contexts to which it can be applied, one attempt to characterise a problem as a big data challenge is through the use of the 'four Vs': volume, velocity, variety and veracity. Volume and velocity refer to the large quantities of data generated and the rate at which it is acquired, respectively – high content screening for drug discovery is a good example of this. Variety refers to the different categories of data that can be used, an example being the ability to acquire bioimaging data at varying resolutions, together with associated metadata and annotations of objects in a typical study. The fourth V, veracity, is an indication of the worthiness of data. While at the early stages of the big data 'hype' only the first three Vs were the main focus, this additional V is particularly important because having large volumes of images that can be acquired at high throughputs is not of much translational value if the data is incorrectly acquired, containing high degrees of noise, scientifically inconsistent or incorrectly processed.

Questions such as 'How big is large volume?', or 'How noisy is an image of good veracity?' remain largely subjective and context dependent. For the purposes of this chapter, we take the rather generic assumption that *big bioimaging data* refers to datasets that are too large (in both number and size) and complex (multiple channels, dimensions, annotations, metadata) to be handled and analysed using conventional methods, which are often largely labour intensive.

In life sciences, one subfield that is often used as a solid example of successful application of informatics to handle big data, in terms of both methodology and software applications, is genomics. It may therefore be argued that the requirements of big data image informatics systems can be met by using technologies that were used successfully in other areas of modern biology, particularly genomics. Note, however, that there are fundamental differences between typical datasets used in imaging and genomics. As discussed by Goldberg et al. [7], in genomics, knowledge of the type of sequencer that was used to determine the DNA sequence AGCTC… is not important information for sequence interpretation. Additionally, the result generated upon sequencing usually suffices as a means of 'quantifying' the sample. No further processing is required to know the sequence. This is, however, not the case in imaging. An image of a cellular structure can only be truly understood within the context of which cell is of interest, in addition to the knowledge of the preparation steps needed before imaging the sample (e.g. choice of fluorescent markers used). Other layers of complexity are then added, such as noise characteristics of the microscope used for acquisition, before an image could then finally be processed and quantified. Thus the needs of the microscopy imaging community are not met by conventional genomics tools, and the problem of big data in microscopy should be treated as a separate issue.

In this regard, the primary aim of this chapter is to provide a review of a number of powerful software technologies (both open-source and commercially available) that the author enjoyed using for microscopy image quantification: ImageJ (1.50i), Fiji (on Java 8), CellProfiler (2.1.1), Icy (1.7.3.0), Imaris (8.1) and Definiens (2.4.2). This is an effort towards assisting life scientists who may be new to bio-image analysis and thus unfamiliar with the reviewed technologies. Due to the dynamic nature of the field, newer versions may be available by the time this review appears in print, and readers are advised to refer to the websites cited for latest updates. Additionally, throughout the discussion, we tend to elaborate on commonly used technical terminology and clarify their relevance to bio-imaging applications; this is a step towards bridging the gap between life scientists and technology professionals in the imaging field.

9.3 THE OPEN-SOURCE BIOIMAGE INFORMATICS COMMUNITY

9.3.1 *ImageJ for Small-Scale Projects*

One of the most popular imaging software packages within the life sciences, as well as the wider imaging domain, is ImageJ [16]. For the past two decades, it has been considered by the community to be one of the pioneering tools for the visualisation and analysis of scientific images. ImageJ was developed in the NIH research centre by Wayne Rasband and is based on the Java programming language [6].

One way to think of a programming language is as a structured set of methods that technology professionals can use to communicate with, and control the behaviour of, a computer. The Java programming language was introduced in the 1990s by Sun Microsystems – a technology giant of the time. One motivation behind introducing Java was the idea of having a platform-agnostic means of developing software, in other words, being able to develop software that can run on Windows, Mac or Linux, without the need to rewrite the code base. Due to this key feature of Java, ImageJ is able to run across various independent platforms.

One of the reasons behind the popularity of ImageJ within the scientific community is perhaps the strong means by which the users can contribute to the software ecosystem, through the so-called *macros* and *plugins*. Macros are simple, custom programming scripts that can automate tasks performed in ImageJ, using ImageJ's macro language. The use of macros requires a very straightforward programming syntax, meaning that users from non-technical backgrounds who are interested in automating relatively complex tasks in ImageJ, can do so without barriers such as the lack of formal programming training. ImageJ also provides a very useful functionality through which the user can record ImageJ actions as macro commands, thus further simplifying the process of generating macros by non-technical users. Plugins, on the other hand, are relatively more sophisticated to develop and can be thought of as extensions to the core ImageJ software. Most of ImageJ's built-in menu commands are, in fact, plugin implementations. Common uses of plugins include filters performing some processing on an image or an image stack, and input/output plugins for reading and writing file formats that are not natively supported.

The concept of a plugin leads us to a particular distribution of ImageJ that is of extensive use within the life sciences field: Fiji [17, 18]. Continuously driven by researchers' requirements, Fiji bundles many ImageJ plugins and offers a range of tutorials and supporting documentation in a Wiki that is continuously maintained and updated. The bioimaging community has benefited from Fiji in terms of centralising and de-fragmenting the plethora of plugins that are continuously being developed and updated. In other words, Fiji can be described as the biologists' 'batteries included' variation of ImageJ. Fiji is also geared towards technical users, as programmers benefit from the flexibility to write code in various languages, in addition to the conventional ImageJ macro language. Figure 9.1 shows how similar the the user-interfaces of ImageJ and Fiji are.

Figure 9.2 shows Fiji's macro recorder and scripting console. While using these two functionalities could go a long way in facilitating automated analysis of bioimaging samples, there is a high degree of manual interaction that is often required with conventional macro scripting, as

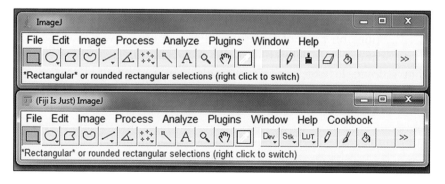

Figure 9.1 User interface of ImageJ and Fiji.

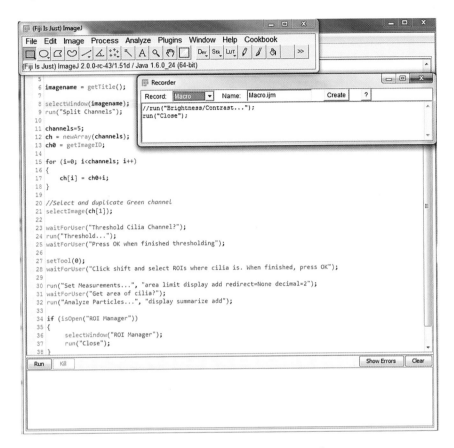

Figure 9.2 Fiji macro scripting console.

it is mainly aimed towards the analysis of one image at a time. Thus, a tool that could be quite handy when trying to automate processes on large datasets is the Batch Process tool.

The Batch Process functionality provides the ability to perform the same actions on a set of images. As shown in Figure 9.3, the dialogue allows the user to define a directory that comprises all your images that you would like to process. Similarly, users can specify an output folder directory, for keeping any generated images that result from processing. Use of this tool ensures that the set of macro commands are applied to every image in the input directory, and it is one way to handle relatively large datasets, where constant manual clicking is not a scalable option.

While Fiji (or more generally, ImageJ) is an effective choice for automating common and relatively complex image analysis tasks, the available

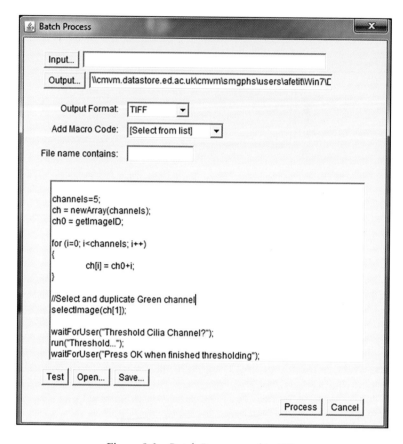

Figure 9.3 Batch Process tool in Fiji.

options for batch processing often introduce a bottleneck when the user wishes to deal with hundreds of samples. In other words, although automating the analysis of multiple images is possible, most of ImageJ/Fiji functionalities are geared towards individual analyses; Scalability issues tend to arise in settings such as high-content screening, where it is possible to generate thousands of high-resolution images per day. In terms of three-dimensional and time-lapse samples, there is a wide range of powerful plugins that exist for the visualisation of such data (e.g. ImageJ 3D viewer), but to the best of the author's knowledge at the time of writing, analysing and quantifying such datasets on Fiji remains a challenge that the bioimaging community faces.

9.3.2 CellProfiler, Large-Scale Projects and the Need for Complex Infrastructure

Motivated by the ongoing needs for biologists to automate the analysis of heterogeneous datasets, particularly for high-throughput applications, the CellProfiler project [10] was initially introduced in 2005 as an open-source means of addressing this. The software was first released by scientists from the Whitehead Institute for Biomedical Research and MIT, and was originally written in MATLAB. MATLAB [19] is a powerful pro-gramming environment that is extensively used by scientists and engi-neers for modelling complex systems. It is a popular choice for developing prototypes and scripts in research settings. However, unlike the Java programming language mentioned earlier, MATLAB is a pro-prietary product developed by the company MathWorks, hence licens-ing regulations apply for developing prototypes in MATLAB.

The CellProfiler project subsequently re-engineered the software to be more open, robust and user-friendly, with new algorithms and features to facilitate high-throughput work. It is currently being developed at the Broad Institute of MIT and Harvard. The new version, CellProfiler 2.0, has moved beyond MATLAB and is now written in Python. The new release aimed to move beyond small-scale projects, highlighting the trend toward quantify-ing imaging data regardless of experiment size [11]. The move towards Python, an open-source language, has enabled the developers to use high-performance scientific software libraries, such as NumPy and SciPy.

Since a key focus of CellProfiler is to ease the analysis of high-through-put images, one powerful feature of the software is the ability to rapidly query images, such as by using keywords, and to analyse the images that fall within a certain criterion in a structured way. The example in Figure 9.4 illustrates a scenario where this is applicable. The researcher in this case was interested in quantifying data from two sources: DAPI and GFP images. Using certain keywords that are associated with the images (in this case, C1 and C2), we could enable CellProfiler to separate DAPI images from GFP, and to rename the results of the query in a way that enables easier identification of the channels (e.g. OrigBlue and OrigGreen).

As a means of making the software extensible and appropriate for large-scale high-throughput settings, CellProfiler was designed to be suitable for integration within high-performance computing (HPC) cluster environ-ments. HPC clusters are becoming increasingly popular in scientific research settings, and are key for long-term adoption of big data analysis

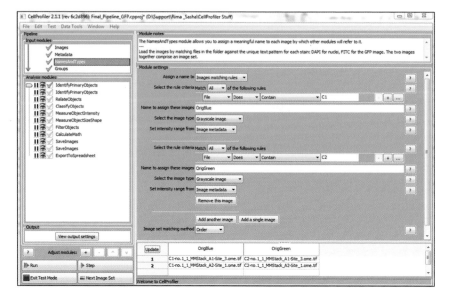

Figure 9.4 User interface of CellProfiler. It is possible to rapidly query images of interest using keywords.

systems. HPCs can be thought of as an aggregation of computing power that can deliver more powerful performance than a single workstation, in order to solve large analytical problems such as simulation, modelling and rendering, the type of problems that a single computer workstation would be unable to address within a reasonable time frame.

In an HPC, you essentially have all the elements that can be found in a conventional desktop computer workstation: processors, memory units, operating system – just in rather large quantities. In other words, they are clusters of computers. The number of processors that each of these computers has varies – between one and four processors nowadays – and each processor can have between two and four cores.

Figure 9.5 shows a typical cluster configuration. When a user wishes to use an HPC, they first access what is called a log-in node. It is effectively a gateway for accessing the system. They are not designed for running any heavy computational work, but are mainly where users can carry out simple tasks such as submitting a job, checking on a job status or editing files. The compute nodes, on the other hand, act as the workhorse of an HPC. When a job is being sent for processing – either parallel or serial – on a cluster, the compute nodes are what is being used. Once the user has accessed the log-in node of an HPC, and upon submitting a job for processing, the job waits in a queue until it is executed. The time

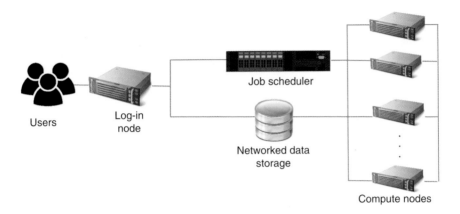

Figure 9.5   A typical cluster configuration.

frame spent in the queue depends on a variety of factors. Management of job queues is carried out by what is called a job scheduler. It is a fairly complex process, but generally it relies on information such as job size, priority, expected execution times (which can be indicated by the user), access permissions (determined by system administrators) and resource availability [12].

With the growing acceptance of open-source technologies in scientific research settings, using a large number of HPC systems requires, at minimum, some familiarity with the Linux operating system. When using Linux cluster environments, the experience is going to be fundamentally different from using, for example, a Windows desktop computer, because there is no use of graphical user interface. In other words, no icons, menus, buttons, etc., so the user interacts with the system by typing commands into a console (Figure 9.6). Knowledge of basic Linux commands is necessary to be able to view files, browse between various directories, delete files, submit jobs for processing and inspect the status of a job undergoing processing.

As mentioned earlier, one powerful feature of CellProfiler is its ability to be set up on HPC systems. When you combine the tremendous processing power of an HPC, with the strong algorithmic capabilities of CellProfiler, some interesting results emerge. This setup is a good choice for handling enormous datasets of high-resolution images, typical of those in drug discovery research settings. The technical process of setting up CellProfiler on a Linux HPC is fairly demanding, and requires familiarity with integrating software systems. While diving into deep technical computing details is beyond the scope of this chapter, setting up such analysis framework is of major interest to the open-source bioimaging

Figure 9.6   Linux HPC where the user has accessed a log-in node of a cluster.

community; thus, the rest of this section will comprise technical guide-lines on a number of points that the author found to be useful when setting this up. It is, however, expected that the user has some working knowledge with Linux, Python and the CellProfiler software. Nevertheless, throughout the discussion we aim to clarify technical terminology where possible for the less technical readers.

9.3.3 Technical Notes – Setting Up CellProfiler for Use on a Linux HPC

The guidelines illustrated here make use of Python's virtualenv tool to set up the dependencies for CellProfiler. Short for 'virtual environment', a virtualenv is used to isolate the dependencies required by different projects in separate places, solving the problem of the 'Project A depends on version X whereas Project B depends on version Y' dilemma.

Simply, a dependency is a file or a software library that is required to be present for an installation to be successful. In the case of CellProfiler, the vast majority of required dependencies are what are referred to as python modules. Modules are files consisting of Python code, and are one of the reasons Python has gained popularity in the science and engineering fields: they enable easy reuse of complex algorithms across different applications as packages. One dilemma that is often faced by technology professionals working on more than one projects based on Python is that these different projects may require different versions of the dependencies to be installed on their system. The use of virtualenv alleviates this through isolating different variations of Python as per the developers' requirements.

Once you have logged onto an HPC, and located an area with sufficient storage, the following are the steps to set up CellProfiler's dependencies:

- Obtain virtualenv:
```
wget ttps://pypi.python.org/packages/source/v/
virtualenv/virtualenv-1.10.tar.gz
tar -zxvf virtualenv-1.10.tar.gz
```
- Create a virtual environment (here named myVE) within the virtualenv01.10/myVE directory.
```
cd virtualenv01.10
python virtualenv.py myVE
```
- In order to activate the virtual environment, browse to the myVE directory, and issue this command:
```
source bin/activate
```
- Now, as long as this virtual environment is activated, work relating to Python and Python modules will be using myVE's version. To deactivate, simply issue the command `deactivate`
- Upon activating the virtual environment, you will be able to use the `pip` tool for downloading a large number of the required modules. `Pip` is a package management system that allows users to install and manage software packages written in Python using straightforward commands:
```
pip install cython
pip install numpy
pip install MySQL-python
pip install scipy
pip install zmq
pip install h5py
pip install python-bioformats
```
- One of the major dependencies for CellProfiler is matplotlib. It is, however, possible that you will get an error when trying to install matplotlib through a **pip** command. This may depend on the version of Linux distribution you are working on. When tested, manual installation of matplotlib, version 1.5.1, was successful.
```
wget https://pypi.python.org/packages/8f/f4/
c0c7e81f64d5f4d36e52e393af687f28882c53dcd92441
9d684dc9859f40/matplotlib-1.5.1.tar.gz#md5=f51847
d8692cb63df64cd0bd0304fd20
```

```
tar -zxvf NameOfDownloadedFile.tar.gz
cd matplotlib
python setup.py install
```

- Similarly, scikit-learn may be problematic when using `pip`, depending on the Linux distribution you are using. Manual installation of the package can be done by downloading the files from the project's Git repository:
```
git clone https://github.com/scikit-learn/scikit-
learn.git
cd scikit-learn
python setup.py install
```

- Libxml2-dev and libxslt-dev are two other libraries that are needed before proceeding further. These are, however, C libraries, and will require administrator rights for installation, because setting them up cannot be done in isolation within the Python virtual environment. You will therefore need to liaise with your IT administrators to make sure that these libraries are available for use on the cluster.

- Once the aforementioned C libraries are in place, you would need to load the C compiler that is set up on your cluster, where the two libraries reside. For example (the path will be different):
```
module load igmm/compilers/gcc/4.9.3
```

- Next, manually install Python's lxml package from the project's Git repository. This step will fail if the aforementioned C libraries are not available:
```
git clone https://github.com/lxml/lxml.git
cd lxml
python setup.py install
```

- It is possible that, at this stage, the dependency prokaryote might be already installed. If you have not explicitly installed it, it is worth uninstalling and then reinstalling it, as it is probable that what is available is only a partial installation, which may result in errors.
```
pip uninstall prokaryote
pip install prokaryote
```

- Finally, CellProfiler can now be installed:
```
git clone https://github.com/CellProfiler/
CellProfiler.git
cd CellProfiler
python setup.py install
pip install --editable
```

- The use of CellProfiler to run jobs on an HPC is best done through what is called headless mode. To run the software headless means to make it work without the use of a graphical user interface, which is suitable for a Linux-based command line environment. Using CellProfiler through the command line requires the use of switches, which are commands that define certain settings when sending a job for analysis. For example, the use of the switch -c directs the software to run without a user interface; the switch -p specifies the pipeline that needs to be run. --data-file provides a location to a csv file that contains information regarding the data that needs to be loaded for analysis. -o sets the default output directory that the analysis results should go to. To illustrate:

```
cellprofiler -r -c -p /home/afetit/humans/
ExampleHumanLoad.cppipe --data-file=/home/afetit/
humans/Input_file_list2.csv -o /home/afetit/output4 -L
DEBUG
```

This will run CellProfiler in batch mode, without a GUI, using the pipeline 'ExampleHumanLoad.cppipe', and will use the images specified in the csv file 'Input_file_list2.csv' as input, storing the results in a directory called 'output4'.

- Note that best practice guidelines for using CellProfiler on an HPC are available on the project's Git repository [13].

- It is worth noting that it is also possible to distribute the virtual environment for other users of the cluster to use, without the hassle of re-downloading and integrating all the dependencies. The new user can do this by downloading virtualenv and moving a copy of the environment (in this case, myVE) to their workspace. Before using CellProfiler, however, it is necessary to create a new virtual environment with the same name (myVE). This will not erase the original contents of myVE, but will only update any links that the environment uses so that they can work for the new user.

- A common error that you may face during the process of setting up CellProfiler is a 'disk quota exceeded' message. This tends to arise when users attempt to set up the virtual environment on the storage space provided on the log-in node of the cluster, which is not designed for storing large files or carrying out computationally intensive processing. The best approach would be to carry out the aforementioned steps on a designated storage area that is linked to the HPC, which you can find out about from your IT administrators.

9.3.4 *Icy, Towards Reproducible Image Informatics*

Another community-oriented project that has gained traction in recent years is Icy, which is developed by Institut Pasteur in Paris, France. The Icy project is motivated by the concept of reproducible research, which encourages authors of publications to provide access to all the information that led to the reported results, so that experiments can be independently repeated and validated by other researchers. Additionally, Icy proposes the concept of extended reproducible research [14], which on top of the original reproducibility specifications should require:

• reusability and extension of algorithms
• standardisation and interoperability of software tools.

Icy aims to achieve this through the use of modular components for carrying out image analysis workflows, through what is called a visual programming framework (Figure 9.7). Visual programming is essentially a

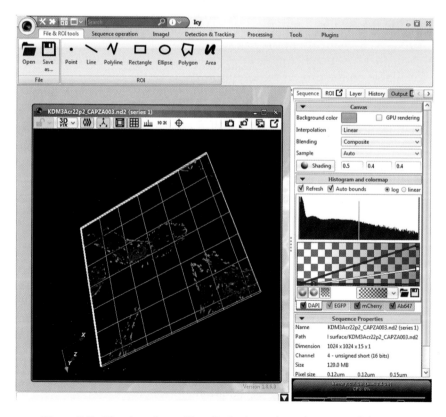

Figure 9.7 User interface of Icy displaying a three-dimensional dataset.

way of allowing the users to create programs by manipulating high-level program elements graphically, rather than by specifying them textually. In other words, different processing, visualisation and analysis modules could be chosen, interconnected in a pipeline and altered to match the user's requirements without directly specifying commands in, say, the Java programming language. This approach aims to make sure that algorithms are reusable by a wide variety of users, while remaining amendable by users from limited technical backgrounds. The project also ensures the availability of a content-management system to handle such tools for users to be continuously involved and engaged, a similar concept to Fiji's plugins update site. This enables users to easily contribute to the wider Icy community through a webpage that is dedicated for each project.

Similar to Fiji, Icy is bundled with a set of tools that have been developed by the Icy project as well as by the wider community. There are more than 100 plugins available for use within Icy [14]. Interestingly, Icy also provides a fully functioning version of ImageJ together with the software, directly available for users to integrate within their workflows. In terms of usability, one powerful feature of Icy is its straightforward user interface, which was designed to make sure that users would already be familiar with the environment and could thus intuitively navigate through it. For example, the main toolbar where users can access most functionalities has a very similar design to the popular Microsoft Office Suite, thus giving the users an intuitive user experience and familiarity with the software. Icy also allows powerful visualisation of 3D images, as well as time-lapse datasets, which are becoming increasingly popular in life sciences.

9.4 COMMERCIAL SOLUTIONS FOR BIOIMAGE INFORMATICS

9.4.1 *Imaris Bitplane*

Imaris [20] is a commercial application that is commonly used in life-sciences, which focuses on the visualisation and analysis of three-dimensional images and time-lapse datasets. A key strength of Imaris is seamless rendering of multidimensional (spatial, multichannel and temporal, x,y,z,c,t) datasets, and the ability to present them in a way that users find easy to interact with. Handling images and time-lapse datasets that are exceptionally large in size can be done seamlessly in Imaris, which scales well when trying to render multiple images of interest.

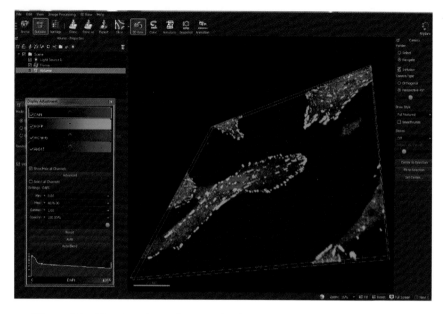

Figure 9.8   User interface of Imaris displaying a three-dimensional dataset.

Another key point about Imaris is its ability to provide reliable particle tracking algorithms, which can be rapidly applied on time-series datasets for a wide range of applications (Figure 9.8).

Imaris is geared towards facilitating the analysis of three-dimensional datasets. The visualisation, segmentation and quantification of such data is more mathematically rigorous and difficult for novice users with limited experience of advanced geometry to programme, hence Imaris is widely used in the bio-imaging field. Thorough documentation and online help videos are available to aid users new to Imaris.

When handling multiple images, it is possible for the user to define certain settings on an image, and then apply the desired settings onto other images that need to be analysed, which can be useful in relatively large studies. Imaris also has the ability to quantify 3D fluorescent data generated from a wide range of sources. Imaris can also produce engaging dyanmic visualisations of datasets for use in publication or presentation.

9.4.2 *Definiens and Using Machine-Learning on Complex Datasets*

Definiens [4] is commercial product that offers a variety of software packages that can help biologists and imaging scientists to rapidly analyse large, complex datasets. One popular software is Tissue Studio,

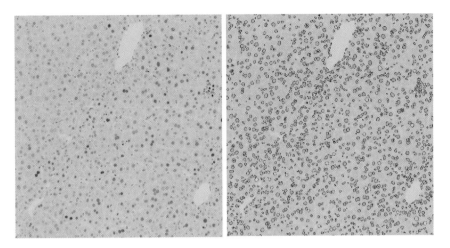

Figure 9.9   Example of how Definiens could be used to rapidly classify different types of cells on histology images. On the left is the original scan, and the rightimage shows how Definiens has classified the cells into two classes.

which aims to facilitate the quantification of histology datasets (Figure 9.9). Unlike the case of fluorescent images, histology images are usually acquired using standard colour camera systems through white light illumination, without the need to excite fluorophores. Consequently, histology data is generally used to get a high-level view of a sample, and is widely used in clinical applications. Additionally, it is common to capture histology images under various resolutions, and to include all of them under one imaging file, giving the researcher the ability to rapidly zoom in and out of areas of interest and inspect their characteristics at high quality. This tends to result in histology datasets being exceptionally large compared to images acquired using fluorescent microscopy systems. Thus, the analysis of histology datasets tends to be treated as a unique domain.

The key strength of Definiens is the ability to rapidly use machine learning algorithms for accurate classification of tissues, cells and subcellular structures (Figure 9.9). Broadly, machine learning is a branch of artificial intelligence that provides computer systems the ability to capture patterns and learn from data, without explicitly being programmed. The way this is applied in Definiens is usually through what is called supervised classification, where the user of the software 'teaches' Definiens on a number of images (or even limited regions of an image) the different categories that they are interested in. For example, if the interest is to differentiate between cancerous and healthy cells, the user assigns a number of cells that they know belong to the *cancerous* and *non cancerous* classes. Definiens' underlying algorithms then capture various patterns from the specified training data, such as colour, morphology and texture, via various sophisticated mathematical formulae. This knowledge is then used to identify other unseen images, thus enabling researchers to rapidly quantify large, heterogeneous datasets.

An additional advantage of using Definiens in research settings is the software's use of server technology, which allows the analysis of large datasets to be done rapidly using a dedicated server, without being limited to the resources of a local computer. One workflow in which this can be applied is where the researcher holds their datasets in a network share that is accessible by Definiens. The researcher then trains the pipeline to pick up patterns of interest on a small subset of the images, and then asks the server to process the entire dataset using the learned patterns. This then frees the user from sitting by the workstation until the analysis is completed; the processed results are returned to the network share that is accessible by the user.

One challenge in using machine learning based commercial solutions is the difficulty to answer questions such as "on what basis did the software classify a particular set of cells as cancerous?" This is, in a way, counterbalanced by the software's robust algorithms, which are often quite precise, when trained accurately on high-quality training sets. Also, implementation of the server requires continuous liaison with the IT support staff and setting certain policies that ensure that the process does not lead to unexpected issues. For instance, the fact that the server can write data back to a network share – which may be heavily used by a research group to hold their images and documents – introduces a layer of complexity and potential risk, although it can perhaps be mitigated by only allowing the server access to storage areas that rely on Definiens for their analyses.

It is worth noting that it is now possible in Definiens to manage job priorities, and assign certain jobs as being more important than others. This tends to be important in scenarios where there are various research groups relying on using Definiens, but some jobs would require longer processing times than others. Note that it is currently only possible to have one copy of the software running at a particular time for a given licence, which can challenge the software's implementation for use between multiple research groups.

9.5 SUMMARY

The big data phenomenon is undoubtedly migrating to affect all aspects of our lives, and bioimaging is no exception. This chapter presented an overview of open-source and commercial software technologies that can be used for handling bioimaging datasets, which are continuously increasing in richness and complexity.

There are various open-source projects that aim to provide resilient image informatics systems for big data applications.

For quantification, there is a wide variety of options that can be employed, and the choice is usually context-dependent and varies largely on the biological question in hand. A number of factors need to be considered before choosing a software platform, and the number of datasets that need to be analysed is perhaps the most important one.

While a large number of commercial solutions exist, they can be difficult to scale due to licensing regulations. Also, it can be difficult to get an understanding of why certain algorithms in commercial packages behave a particular way as details of the underlying algorithms may not be easily available.

In the author's opinion, one of the main challenges impeding momentum in the area is the lack of a 'common language' that can help life scientists and technical professionals to establish computational frameworks for increasingly heterogeneous imaging datasets. This will certainly give rise to new professions in the future, which sit on the interface between computational biology and computer science, and can help realise this vision. Future efforts in the area should focus on empowering life scientists with the skills needed to quantify and manage their imaging data, and to expose computer scientists to research problems in the bioimaging field, as part of their image processing teaching.

9.6 ACKNOWLEDGMENTS

Many thanks to Semir Kazazic, Ewan McDowall, John Ireland and David Perry, for their continuous advice and support on matters concerning IT infrastructure. Thanks to Sébastien Besson for discussions and advice on using OMERO and the Bio-Formats library. This work was financially supported by the MRC and the BBSRC.

9.7 REFERENCES

1. Kevin W. Eliceiri et al., Biological imaging software tools, Nature Methods 9, 697–710 (2012).
2. David Haumann et al., Big Data in Microscopy: Data analysis and management in complex microscopy experiments, da-cons GmbH (2013).
3. Viktor Mayer-Schonberger and Kenneth Cukier: Big Data: A revolution that will transform how we live, work and think (2013).
4. Definiens, Available from: http://www.definiens.com/
5. Bioimage analysis software: is there a future beyond ImageJ? Bioimage Analysis Workshop (2012), Available from: http://bigwww.epfl.ch/eurobioimaging/booklet.pdf.
6. Caroline A. Schneider et al., NIH image to ImageJ: 25 years of image analysis, Nature Methods, vol. 9 no. 7, 671–675.
7. Ilya G. Goldberg et al., The open microscopy environment (OME) data model and XML file: open tools for informatics and quantitative analysis in biological imaging, Genome Biology (2005).
8. Bio-Formats, Open Microscopy Environment (OME), Available from: https://www.openmicroscopy.org/bio-formats.
9. Melissa Linkert et al., Metadata matters: access to image data in the real world, The journal of cell biology, vol. 189 no. 5, 772–782 (2010).
10. Anne E Carpenter et al., CellProfiler: image analysis software for identifying and quantifying cell phenotypes, Genome Biology (2006).
11. Lee Kamentsky et al., Improved structure, function and compatibility for CellProfiler: modular high-throughput image analysis software, vol. 27 no. 8, 1179–1180 (2011).
12. Saeed Iqbal et al., Job Scheduling in HPC Clusters, Dell Inc (2005), Available from: http://www.dell.com/downloads/global/power/ps1q05-20040135-fang.pdf.
13. Adapting CellProfiler to LIMS environment, CellProfiler Wiki on Github, Available from: https://github.com/CellProfiler/CellProfiler/wiki/Adapting-CellProfiler-to-a-LIMS-environment.
14. Fabrice de Chaumont et al., Icy: an open bioimage informatics platform for extended reproducible research, Nature Methods 9, 690–696 (2012).
15. Jason R. Swedlow et al., Bioimage informatics for experimental biology, Annual Reviews in Biophysics, 38: 327–346 (2009).
16. ImageJ, Image Processing and Analysis in Java, Available from: https://imagej.nih.gov/ij/index.html
17. Schindelin, J., Arganda-Carreras, I. & Frise, E. et al. (2012), "Fiji: an open-source platform for biological-image analysis", Nature methods 9(7): 676–682, PMID 22743772 (on Google Scholar).
18. Fiji, Available from: http://fiji.sc/#download
19. MATLAB, The MathWorks Inc., Natick, Massacheusetts, United States. Available from: https://www.mathworks.com/products/matlab.html
20. Imaris Image Analysis Software, Available from: http://www.bitplane.com/

10

Presenting and Storing Data for Publication

Ann Wheeler[1] and Sébastien Besson[2]
[1] *Advanced Imaging Resource, MRC-IGMM, University of Edinburgh, UK*
[2] *Centre for Gene Regulation & Expression and Division of Computational Biology, University of Dundee, UK*

This chapter is in two parts, the first describes how data should be presented for publication and the current standards for microscopy image processing and presentation according to leading scientific journals and research funding agencies. The second covers storage of data which will be required prior to generating figures for papers and subsequent to publication. The two sections are separated for ease of reading but are interdependent as a figure cannot be produced or distributed to the scientific community if the data for it is incorrectly stored and documented.

10.1 HOW TO MAKE SCIENTIFIC FIGURES

Scientific figures are made for a variety of purposes, and their means of construction take account of this. At a basic level a rough figure would include a summary of some data for a project lead, lab, group or team meeting. These are often assembled in Microsoft Powerpoint, Prezi

Standard and Super-Resolution Bioimaging Data Analysis: A Primer, First Edition.
Edited by Ann Wheeler and Ricardo Henriques.
© 2018 John Wiley & Sons Ltd. Published 2018 by John Wiley & Sons Ltd.

(https://prezi.com/) or Open Office Present. These types of figures are low resolution. Powerpoint and Prezi are not image manipulation packages and prefer images in JPEG or PNG format. They can accept 8-bit RGB TIFFs which are a flattened image. These packages do not read 16-bit TIFFs, stacks of images including RGB composites which ImageJ makes or 3D projections. All of this data need to be saved in JPEG format. Screenshots of this type of data can be suitable for display.

10.1.1 General Guidelines for Making Any Microscopy Figure

- Use distinct colours with comparable visibility, consider colour-blind individuals by avoiding the use of red and green for contrast. It is well worth recolouring primary data, such as fluorescence images, to colour-safe combinations such as green and magenta, turquoise and red, yellow and blue or other accessible colour palettes. Use of the rainbow colour scale should be avoided.
- Use solid colour for filling objects and avoid hatch patterns.
- Avoid background shading.
- Unusual units or abbreviations should be spelled out in full, or defined in the legend.
- State the resolution/scale of the image by using a scale bar.

For more formal presentations such as the creation of figures for academic degrees (BSc, MSc or PhD/DPhil), journal publication of grant applications different rules apply. Here resolution, correct colour representation and calibration of the images is very important. Before one starts generating the figure it's advisable to find out what the requirements for the figure would be. For instance, some journals have standards for the resolution of figures. Examiners and peer reviewers need to be able to study the scientific figures to establish whether the data supports the hypothesis tested or is correct. Reputable journals have 'instructions for authors' who are considering preparing a manuscript. What is expected in a scientific figure will be written up as part of this.

Basic research journals, such as Cell Press, Nature, Company of Biologists Journals have strict figure guidelines:

- http://www.cell.com/figureguidelines
- http://jcs.biologists.org/content/manuscript-prep#4.1
- http://www.nature.com/nature/authors/submissions/subs/
- http://jcb.rupress.org/site/misc/ifora.xhtm

There are generally maximum figure sizes. Some journals recommend converting the high resolution TIFF figures into JPEGs prior to combining into a PDF for submission. These can be summarised as follows:

- The maximum acceptable figure size is either A4 for European Journals or 'Letter' for USA-based journals.
- Resolution of 300 dpi for colour or 500 dpi for black and white is required.
- For micrographs, use a scale bar to show the magnification and give the length of this in the figure legend.
- Bold uppercase letters (A, B, C, etc.) should be used to distinguish figure panels; Arial font is preferred for figures.
- The optimum font size at 300 dpi resolution is 8 pt.
- Images should be prepared in RGB colour space for ease of production.

Scientific figures can be made in a range of image processing software. However, the standard software used is either Fiji/ImageJ or Adobe Photoshop/Illustrator/Elements. It is possible to generate figures in GIMP and other image adjustment software.

A useful tip for figure making is to duplicate the original images, taking note of their original location, and store them in a figure folder. Presentation of scientific figures is paramount, and images may need to be cropped to size, and an expanded insert may be needed to give the reader a better view of the image data.

ImageJ has a lot of built-in functionality to facilitate generation of figures. Making a stack of the contingent parts of the figure, using the stack–shuffle tool to move frames around, adding a scale bar to the final image in the stack, carrying out background subtraction/intensity adjustments on all of the stack not only speeds figure making but also enables researchers to adhere to good scientific practice as all parts of the figure will be adjusted equivalently. The Image–Stack–Make Montage tool, generally with a border spacing of 5–10 pixels, allows all parts of the figure to be laid out equally and tidily. To ease scientific figure generation for images, an ImageJ plugin called Scientifig (Aigouy, 2013) has also been generated specifically for the generation of figures.

10.1.2 *Do's and Don'ts: Preparation of Figures for Publication*

So what is expected in today's journals? First of all, a concise and clear description of what has been done to the sample from the point where it is prepared to presentation. The fixation, staining and imaging methods (including equipment) should be described. It is acceptable to adjust images for the purpose of quantification. However, particularly if a given channel is used for masking, the procedure needs to be clearly documented. The following need to be included.

- In the section, the type of equipment (microscopes/objective lenses, cameras, detectors, filter model and batch number) and acquisition software used should be specified. Equipment settings for critical measurements should also be listed.
- State the measured resolution at which an image was acquired and any downstream processing or averaging that has enhanced the resolution of the image.
- A single Supplementary Methods file (or part of a larger Methods section) titled 'equipment and settings' should list for each image: acquisition information, including time and space resolution data (x,y,z,t and pixel dimensions); image bit depth; experimental conditions such as temperature and imaging medium; fluorochromes (excitation and emission wavelengths or ranges, filters, dichroic beamsplitters, if any).
- Processing software should be named, with the version and manufacturer. Or if the software is freeware, the link to the website or Github repository with the code clearly stated. Manipulations indicated (such as type of deconvolution, three-dimensional reconstructions, surface and volume rendering, 'gamma changes', filtering, thresholding and projection) must be described.
- Images submitted with a manuscript for review should be minimally processed (for instance, to add arrows to a micrograph). If 'pseudo-colouring' and nonlinear adjustment (e.g. gamma changes) are used, this must be disclosed. Adjustments of individual colour channels are sometimes necessary on 'merged' images, but this should be noted in the figure legend.
- Authors should retain their unprocessed data and metadata files, as editors may request them to aid in manuscript evaluation. If unprocessed data is unavailable, manuscript evaluation may be stalled until the issue is resolved.
- For more analytical papers, a dedicated website, e.g. SIMcheck (http://www.micron.ox.ac.uk/microngroup/software/SIMcheck.html), OpenSPIM (http://openspim.org/) or Multicell3D (http://www.crm.ed.ac.uk/tools/multicell3d) could be used for transparency. If you have an image analysis facility they may put the scripts you use on their website (e.g. http://bigwww.epfl.ch/sage/) but these need to be referred to in the materials text.

- All digitised images submitted with the final revision of the manuscript must be of high quality and have resolutions of at least 300 dpi for colour, 600 dpi for greyscale and 1200 dpi for line art.

10.1.3 *Restoration, Revelation or Manipulation*

A basic rule of thumb is that images should be acquired to be representative of the sample. So if a given structure was either very rarely present or the sample looked incorrect, then to avoid unethical temptations it is best practice not to acquire this image in the first place. More experienced scientists, even not if in the same field will be able to identify sick cells or model organisms, necrotic or poorly prepared tissue or samples that have been poorly prepared, for example where imaging occurred before the mounting medium set and the sample slipped. If in doubt, ask a researcher who is experienced in imaging. The good name of journals and government or charitable funding bodies depends on scientific integrity. Such is their concern over data fraud that they now routinely check their submitted data using bespoke tools to detect fraud (http://scienceimageintegrity.org/). It is also common practice to ask for raw data to be deposited in an online repository for community access. Resources such as Dryad (http://datadryad.org/) or Figshare (https://figshare.com/) can currently be used for this purpose, although projects such as the EuroBioImaging image data repository, when complete, would offer a more specific location for scientific images (http://www.openmicroscopy.org/site/news/image-data-repository-idr-project-update).

There is a fine line between minor adjustments to ease readers understanding of the real data presented and all out data fraud. So where does the line lie? this is a difficult subject which is hotly debated, for more information, see: Yamada, 2004; Blatt, 2013; North, 2006.

If anyone at the journal is in any doubt about the integrity of the data, including those in the production department, the issue is referred to the editors, who will request the original data from the authors for comparison to the prepared figures. If the original data cannot be produced, the manuscript may not be accepted. Any case in which the manipulation affects the interpretation of the data will result in non-acceptance, and any cases of suspected misconduct will be reported to an author's home institution or funding agency.

Scientific fraud by misrepresentation of images can cost someone their scientific career, and rightly so. Data should be as accurate and representative as possible at the time. However, we work in a world of method and technological development. A decade ago, lasers were weaker, detectors less sensitive and many reagents were dirtier or weaker. Even now, there is still much work to do. The sensitive sCMOS detectors for fast image acquisition have fixed noise patterns which do not vary and are not part of a biological or medical experiment. Many detectors create noise, which needs to be removed, as this masks the real data. Some reagents still leave a lot to be desired, antibodies are dirty, the imaging instrument may not be as clean as it could have been, the quantum yield of a given probe may be low. In these cases it is acceptable to adjust images to highlight the data and suppress the noise.

Journals call these reasonable adjustments: background subtractions which remove detector noise; contrast enhancements to reveal weak signals, provided they are visible but still representatively weak; cropping of images, particularly to provide an informative insert. These are acceptable, provided it is clear what has been done and how. These rules apply to the displayed images. For data acquired using quantification of images, the methods used to segment and quantify the images must be clearly described. Use of commercial tools or published open-source resources are recommended here, although it is important to state the manufacturer or author and the software version and to include a citation if at all possible. ImageJ, as an open-source image processing tool, has clear documentation available to the scientific community detailing how the algorithms or scripts implemented in the software work and it is an accepted community standard for image processing. An example from the Nature family of journals is included below.

http://www.nature.com/nmeth/journal/v3/n4/full/nmeth0406-237.html

1. Simple adjustments, applied uniformly, to the entire image are generally acceptable. Changes to brightness, contrast, and colour balance fall into this category because they affect the image in a linear fashion. However, it is not acceptable to adjust brightness or contrast levels to such an extent that image data are truncated or lost (giving a white or black background). Such changes may give a clearer picture of bands which are 'of interest' in a gel, but they will mask background, including information that is important for quantification and validation. We will not accept image data that are processed in this way.

2. Cropping and resizing an image is usually acceptable, but both may on occasion be construed as inappropriate manipulation. If cropping, ask whether your motivation is to improve the composition of the image or to hide something that complicates interpretation. The former reason is acceptable; the latter is not.

3. Digital filtering of an image is not encouraged because it can easily mask important information. Most filters use mathematical functions that are nonlinear. There are circumstances in which digital filtering is a necessary part of the experimental methodology. If so, filter processing must be clearly justified and documented in the figure legend or under Methods. Such documentation should include reference to the software version and specification of the filters and any special settings that were used.

4. Combining images is acceptable only if it is clear to the reader that the images are from separate sources.

5. Selective alteration or processing of one region of an image is not acceptable. Such manipulations include 'cloning' or copying objects or sections within or between images and 'smudging', blurring, blending, and other manipulations that are applied locally within an image. Common examples involve sections of an image that have been cloned or blended to clean up a dirty preparation or to mask an unwanted blemish. Such manipulations constitute inappropriate handling at best and are unethical. If the data require such processing, repeat the experiment.

6. When comparing digital images, it is important that each has been acquired under identical conditions, and any post acquisition image processing must be applied identically. If the background or colour balance must be adjusted among images within a group, this must be acknowledged in the figure legend or under Methods. Quantitative analysis of images should always be performed on uniformly processed image data, and the data should be calibrated to a known standard. Most instruments, including fluorescent microscopes, are prone to fluctuations and drift over time, so it is advisable to include appropriate internal standards as checks against such changes.

7. Image data should be documented both with representative images as well as with quantitative statistical analysis of sufficient numbers of experiments. It should be self-evident that experiments that include image data should be repeated and the data analysed for significance. We expect conclusions drawn from image data to be justified based on their quantitative assessment, not on anecdotal observations.

To summarise all of this, a figure should clearly show the biological phenomena. If the structure is too small to see in the image, an extra panel expanding the feature should be included. Images should always be scaled and clearly labelled so the reader can determine the features of interest. Figure 10.1 shows an illustrative example of a figure considered to be acceptable for publication which was generated in ImageJ.

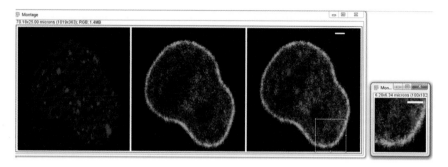

Figure 10.1 representative example of a scientific figure which is acceptable for publication. A super-resolution (STORM) image of a HeLa cell nucleus labelled with antibodies to HP1 alpha (red) and LaminB (green). Scale bar indicated 2 microns. The high resolution insert, right, is indicated by a yellow box on the merged panel.

10.2 PRESENTING, DOCUMENTING AND STORING BIOIMAGE DATA

As is clear from the policies of journals and funding agencies, the presentation and storage of bioimage data go hand in hand.

> Raw image data must be saved and archived intact and without alteration as part of good laboratory practice. Processing of digital images should be done on a copy of the image data file, not on the original. Retaining raw image data is important because they serve as the standard against which the final image can be compared, and they ensure a route for recovery should a mistake be made during processing. We recommend that image data be saved in TIF format. JPEG compression affects the resolution of the image, and information is lost in the process of conversion.
>
> Nature Journals

The proof of a scientific experiment is that it can be recapitulated by others. For this, details of how the experiment was done need to be described to those who may wish to further the presented studies. To present microscopy image data for publication it is necessary to state in the methods and/or supplementary sections of papers:

- how the image was generated, including the specification of the instrument
- what the scale of the image is
- what has happened to the image, if it was processed and if so how
- which statistics were used and in what package
- how the images were prepared.

For large studies, retaining this information with images at the point of capture is extremely useful.

To meet these transparency criteria the first matter to be addressed in preparing images for publication is that the analysed image needs to be stored in a format that can always be opened, not only by the researcher but also by the PI and others involved in the project. Reputable scientific journals will require access to the raw data included in a study for peer review. Journals and funding agencies, for good data governance, require storage and access of data for up to 10 years post publication. Therefore it is important that this is possible, and that data is stored in such a way to enable this. Further, journals may not want data stored in a proprietary format that may not be accessible.

Historically the TIFF format was the preferred method for storage of image data as it retains the numerical matrix which the image is made up of in its original format. TIFF still has the advantage of being a long-standing file format which is opened by many image processing and analysis applications, particularly open-source programs.

10.2.1 *Metadata Matters*

Metadata acquired when an image was captured is information about the image, for example which instrument acquired the image and what settings were used for image acquisition. Most manufacturers' proprietary data formats record this valuable information with the image, hence the range of microscopy image data formats. For publication it is essential to include: objective used, laser power, filtersets used and a host of other information about the data (see above). However, recording all of this information at the point of image acquisition can become arduous for researchers. There is, therefore, an incentive to use manufacturers' propriety data formats when acquiring images as essentially the hard work of recording the instrument settings when the image was acquired has been done as part of image capture. This data is generally stored in a format that is proprietary to the imaging system vendor. The format will generally not open in competitors' software and sometimes not in third-party analysis software either.

Many file formats proprietary to microscope manufacturers are based on tagged image file format (TIFF) (AwareSystems, n.d.) in the way that they store the image information. However, issues tend to arise in the ways the metadata is handled. This is caused in part by the TIFF format that allows for a flexible set of information fields. Unfortunately there is a lack of an agreed-upon standardised way of storing metadata in TIFFs and a lack of standard tools for opening data. Hence manufacturers tend to define their own metadata formats, which are often incompatible with competitors' formats. This can be difficult if two imaging modalities from competing companies are used in a study and their data needs to be compared or correlated. Most commercial microscope manufacturers will update either their equipment or software regularly to ensure that their products remain competitive. A commercial vendor's major version upgrade of software can provide issues with reading of data from older versions or between differing bioimage capture systems. Further, most imaging hardware is supported by commercial ventures and therefore is subject to the whims of the market. Microscope and image analysis vendors can go out of business, be taken over or discontinue a product. If this happens during the course of a study it can pose problems. How does one open and analyse data if the original piece of software is discontinued and isn't compatible with the image analysis computer's operating system?

10.2.2 The Open Microscopy Project

As a move towards addressing this, the OME Consortium has developed a specification that covers most of the metadata in proprietary formats from a wide variety of sources; centred on imaging metadata in the life sciences domain. It provides a straightforward format for microscope imaging data that can be used by software manufacturers working with TIFF (Linkert, 2010).

OME-TIFF is a multiplane TIFF file that contains this image metadata in the OME-XML header. The parameters of the OME and TIFF can include: XYZ dimensions of the image and pixel size, and extensive metadata on image acquisition, objective, filtersets, exposure time, any image annotations and regions of interest (ROIs). The OME common specification is essential for the exchange of image data between different software packages. The OME-TIFF includes OME-XML, in which the XML stores metadata according to the OME model, serving as both a standalone file format and a convenient 'holder' for metadata acquired with an image, which can be migrated from one site or user to another. This allows the pixels to be read with any TIFF-compatible program, and the metadata to be extracted with any OME-aware application. Recent increases in image data size due to applications such as high content screening, super-resolution imaging and light sheet microscopy have led to the OME-TIFF being developed for compatibility with several different data containers for high performance storage of data, BigTIFF being an example of this. The OME project was initially published in 2005 (Goldberg, 2005) and has been in expanded since to follow any new developments in the imaging field.

10.2.3 OME and Bio-Formats, Supporting Interoperability in Bioimaging Data

One of the most powerful technologies in the bioimaging domain that is designed for handling complex data is Bio-Formats. The Bio-Formats project was created by the Open Microscopy Environment to address the issue of data interoperability (Linkert, 2010). Bio-Formats is a community-driven project with development teams at LOCI at the University of Wisconsin-Madison, University of Dundee and Glencoe Software. It comprises a standardised application interface for reading and writing image data using standardised, open formats. At time of writing, Bio-Formats can be integrated into open-source analysis programs such as ImageJ, CellProfiler and Icy, informatics solutions like OMERO and the JCB DataViewer, and commercial programs like MATLAB. The current release of Bio-Formats 5, improves support for high content screening, time lapse imaging, digital pathology and other complex multidimensional image formats, reading and converting over 140 file formats including file formats generated by systems from major microscope system manufacturers to the OME-TIFF data standard. Essentially, Bio-Formats enables the reading of a wide variety of proprietary file format and supports output to the aforementioned OME-TIFF specification.

This issue is of tremendous importance when handling imaging data, and is rapidly evolving due to ongoing advances in acquisition systems. The research cycle in certain disciplines can be long, cell biology being a particular example, and sometimes the software version that the data was acquired with is updated and the older data is un-openable in the newer software versions. Groups move, research equipment changes and manufacturers can provide major version updates in the 3–5 years from initial data acquisition to publication. The research governance guidelines for UK and EU funding agencies require data to be available for 10 years post publication. This can mean between 12 and 17 years after acquisition. Bio-Formats gives researchers the flexibility to allow researchers to choose whether to leave data in the proprietary file format generated by a manufacture or convert all the images into the interoperable OME-TIFF file format. Certainly, without tools like Bio-Formats, users would constantly be confined to using commercial software developed by the manufacturers, and would be prevented from easily exploring the plethora of visualisation and analysis software provided by the open-source community.

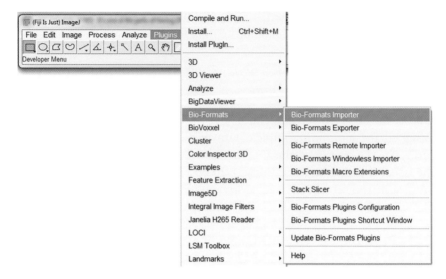

Figure 10.2   Using Fiji/ImageJ to open data using the Bio-Formats importer.

Bio-Formats is routinely updated by the developers as new file formats and applications are continually developed. In ImageJ this can be automated through Fiji updates sites, or updated directly through the plugin (Figure 10.2).

When opening an image or stack of images in Bio-Formats considerable power is given to the end user. Data can be organised, dimensions swapped, tiles stitched, metadata viewed, and channels or stacks split at the point of opening. Data can be opened in a 'virtual' format where only the first image in a stack is read by the machine's RAM (useful for large data files) (Figure 10.3).

The Bio-Formats project is at present an invaluable community resource, enabling not only easy access but fast adjustments of data speeding the path from acquisition to publication. More information about the project can be found here: http://www.openmicroscopy.org/site/support/bio-formats/

10.2.4 *Long-Term Data Storage*

Funding guidelines state that original data must be kept for quite some time, often between 5 and 10 years after the data has been published. For the project lead, this can prove to be very challenging. Collecting sufficient data for a scientific study can take anything between 2 and 10 years. Although some are lucky and make it through peer reviewed

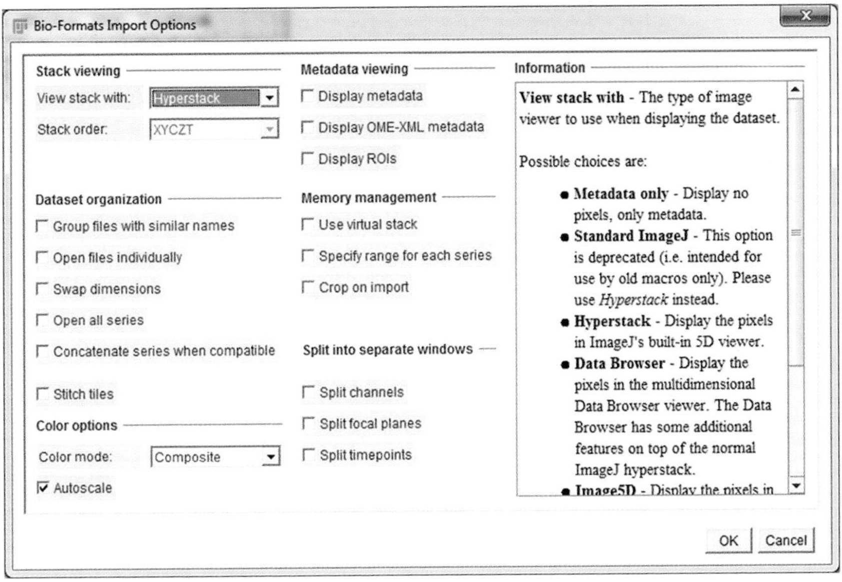

Figure 10.3 The Bio-Formats importer dialogue box.

publication in a few weeks, currently most groups find that the cycle of submission, addressing reviewers comments, resubmission and editing of images can take up to a year. Once the paper is accepted, the clock then starts ticking for data retention! The lifecycle of scientific data can be 15 years or more. Imaging data keeps on getting bigger: for 2D single coloured images consisting of a 512×512 pixel array, a TIFF file size can be as small as 700 Kb. However, for a single experiment which requires 100 data points, such as cells, organoids or tissue images with 3–4 parameters measured in different fluorescent channels, the data size increases maybe to 100 Mb. Several experiments need to be done. Live imaging and 3D imaging present the next increase in data magnitude. Live imaging may consist of 100 multichannel images per field measured and 24 fields may need to be measured, so an expected data size could be a few GB. 3D imaging is similar because a volume will need to be sampled at Nyquist resolution to collect all the data, which might commonly mean 50–100 multicolour slices in a Z-stack. New methods such as high content screening where several thousand images are generated mean that a project using new imaging modalities may generate data in the terabyte range.

Most studies have several different people making important contributions to the imaging data. For the project manager, often a busy lecturer or professor, with classes to teach, committees to attend and other administrative burdens, keeping on top of data can be challenging. A good data management plan, which has been appropriately costed and considered is a hallmark of an excellent scientific study.

10.2.5 USB Drives Friend or Foe?

USB and external hard drives are extremely useful for moving data around. They are capable of storage from 4 Gb to 2 Tb of data and are agnostic to the data type. For fast data transfer USB 3 drives present a 10× improvement in file transfer speed. Using external storage media is not, in my experience, a good strategy for long-term data storage. USB drives can get lost, data can be erased by accident or a virus can be introduced onto the disk. Furthermore, hard disks have a finite life span, often as little as 1–5 years. There are circumstances where the only available option is to store data on an external hard-drive or indeed an internal hard-drive in a computer. In this case, it is essential that an identical backup is made. Image data can often be difficult to collect, perhaps generated from genetically modified model organisms or particularly difficult to grow cells. These experiments are costly to repeat and sometimes impossible if a particularly technically experienced PhD, postdoc or staff scientist has carried out that work and then left. The cost of an additional external drive is often less than £100 or $100. Good backup software is available for free. Websites such as www.techradar.com and cnet.com will have advice and free downloads of backup software available.

Backups can be made to other external disks, Blu-Ray or tape backup is still the standard in many research institutions, but this requires careful cataloguing and record keeping on the part of the data curator and the data generator.

10.2.6 Beyond the (USB) Drive Limit

Not everyone is fortunate enough to have a cutting edge scientific computing department to support their imaging research work, but it is very strongly advisable to engage them at the earliest opportunity in a complex imaging study, because their advice will be invaluable and may prevent hours of lost time. If this help isn't easily available then generating a simple computer network is not too complex a task. A network associated storage (NAS) device can be used to store imaging data. These devices use a protocol called RAID (redundant array of independent disks) to divide data between several external hard disks, https://en.wikipedia. org/wiki/RAID. RAIDed storage combines multiple physical hard drives into one logical unit. RAID1 or higher means that data is mirrored on different physical disks, so if one disk fails it can be removed and replaced without compromising the data. This provides a fairly inexpensive method of storing data for one or many imaging projects as multiples of 2 Tb external disks can be used. NAS units come with built-in software which allows incorporation into wired and wireless networks and can also communicate with backup devices. This provides an advantage over an individual external hard drive because disk failure does not necessarily mean loss of data. However, the device is still a physical unit, a power surge, mishandling such as dropping, and multiple simultaneous disk failures could still mean data failure. An NAS can be backed up onto another NAS or onto a tape drive.

10.2.7 *Servers and Storage Area Networks*

Storage area networks (SANs) are networks of computers or servers, controlled by a network switch. For universities and research institutes, SANs are the main way that data is managed. The SAN will have its own array of networked storage devices which are not accessible to end users so they can back themselves up (Figure 10.4).

SANs have several benefits for image analysis. First of all they can be connected to many different computers or servers using a local area

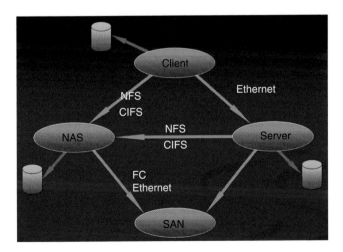

Figure 10.4 A schematic diagram of a storage area network https://en.wikipedia. org/wiki/Storage_area_network.

network (LAN). From an imaging perspective, the computers controlling the microscopes, scanners and cameras that acquire image data can all be connected together through a LAN. The SAN can also 'see' servers, which is useful because many high content imaging systems or high spec imaging systems generate large volumes of data which is stored or locally processed on a server. Which computers are part of the LAN and what access rights they have (e.g. read data, write data, edit/modify data) can be controlled by the network administrator. The network administrator does need experience to do this and it is often preferable to ask IT or scientific computing to do this job for you. Often a SAN will already exist that can be modified to address imaging needs. Access rights can be controlled through a managed infrastructure. This requires a systems administrator. Generally the IT services can provide such a service and work with principal investigators or project leads to manage access and backup of data.

Cloud-based systems are a virtualised version of a SAN. Instead of the servers hosting data being physically present in the workplace, such as a university or research institute, the data is stored elsewhere. Facilities such as Dropbox (www.dropbox.com), Google Docs, Amazon or specialist products for storing image data such as ACQUIFER (https://www. acquifer.de/) and other hosting providers offer such a service. Generally, the first few terabytes or storage will be free but beyond this, storage is paid for. Cloud-based systems have the advantage that they are not tied to one institution, and access rights to the storage will not be dependent on institutional affiliation. It is also possible to harness clouds for data processing as well as storage – freeing up physical computers. This can be advantageous when participating in research projects between different institutions. Collaborators can be invited to view, read or write to given shared folders, and a variety of content including images can be stored in open-source formats such as OME-TIFF, data analysis and annotations can be uploaded. The data isn't generally tagged to enable comparison between datasets, so cataloguing depends on the efficiency of the uploader and the care of all collaborators. For image data there is an immediate problem of scale. Newer imaging technologies generate ever larger data files and these can quickly fill up the free storage, requiring data to be paid for. Further, there is the question of data ownership: the data is being stored by a third party and this may not be as secure as systems in place in host institutions. Finally, if the third party goes out of business, there is the question of what happens to the data and who inherits it. However, these solutions are very useful and are currently widely by many researchers.

10.2.8 OMERO *Scalable Data Management for Biologists*

As mentioned earlier, with the increasing complexity and heterogeneity of bioimaging data, handling and organising images is highly important, because using conventional means of keeping data on unstructured network shares or local workstations is no longer sufficient. A technology that is developed by the OME team which aims to address this issue is *OMERO* (Allan, 2012). OMERO allows life scientists and users of imaging technologies to seamlessly import, archive and manage images, together with their associated metadata and annotations (e.g. ROIs). The OMERO project works towards being a fully collaborative, extensible image management environment, by providing features such as creating user groups that resemble the real life research groups. For example, the principal investigator of a group can be assigned as the group owner, who has full read/write permissions on the group's data, and other collaborators or contributors can be given access to view a subset of images that are relevant to their research, which can be annotated or analysed as desired with the results remaining associated with the images. For further information about this, see: https://www.openmicroscopy.org/site/products/omero.

Implementation of OMERO is a two-step process: setting up an OMERO server system and installing the client application. For the first step, an OMERO server needs to be deployed, and the authors would strongly recommend liaison with your local IT administrators if at all possible for this. The server provides the ability to store the image data, as well as making sure it conforms to the OME specifications. Further information about this can be found at: https://www.openmicroscopy.org/site/support/omero/sysadmins/.

The second step is relatively straightforward and requires downloading an application called OMERO.insight to a workstation (Allan, 2012). OMERO.insight is the front-end client of the OMERO server and can connected to one or many different OMERO servers, provided that the address of the server is available. OMERO.insight can then be used to upload, view and annotate images on the server (Figure 10.5). OMERO has interoperability with ImageJ and MATLAB, which makes it straightforward to integrate image analysis workflows generated using these tools into data archiving routines.

The OMERO.web client can be used to visualise data stored on an instance of an OMERO server if a URL is provided by the OMERO system administrator. The advantage of this is that when an end user

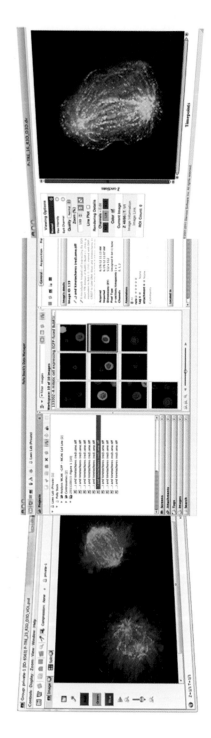

Figure 10.5 The OMERO server enables storage, annotation and analysis of image data in one easy-to-use interface.

logs in, they do not need to have the server information and can still visualise and annotate data. Furthermore, individual files of image data can be shared with selected other users in different groups by adding it to a 'Shares' list. The web client enables data to be visualised and shared between multiple institutions without the requirement for a shared server or IT infrastructure.

OMERO.figure enables generation of publication quality data directly from the OMERO data repository.

Introduction of useful tools such as OMERO.figure makes the process of taking data through to publication seamlessly without resorting to multiple versions of the same image stored in several places more efficient (http://figure.openmicroscopy.org/). OMERO.figure, first released in 2014, combines OMERO's powerful image rendering and metadata to provide a tool for rapid figure creation. In OMERO.figure each figure panel becomes a multi-dimensional image viewer, allowing the end user to zoom and pan, adjust rendering settings and even scroll through Z and Time. The panels can be snapped to a grid to make a perfectly aligned figure which includes regions of interest and scale bars which are seamlessly created from the image metadata. Figures can be exported as TIFF images or PDF documents so the end user can move seamlessly to other editing tools such as Adobe Illustrator. Alternatively OMERO enables the publication the raw data associated with a paper or can provide a link to raw image data using OMERO Figure. Examples of this include JCB Data Viewer or Image data Resource (https://idr.openmicroscopy.org/about/) (Williams, 2017). Taken together the flexible, extensible OMERO workflow makes the processes of carrying out, storing imaging data more convenient and are a great benefit to the community.

10.3 SUMMARY

Image data should be able to be viewed, manipulated and shared as readily as genomic data. The provision of tools for integration of these datasets is still being developed, although we expect it will be fully available in the next few years because considerable community efforts are being made for this. It is hoped that the interested reader, though this chapter, has become acquainted with image data management which is a necessary part of any biomedical research project, and with the current community standards for production of image data containing figures. As the equipment to acquire images advances apace, and datasets become larger and more complex, a proficient knowledge of image data management will become increasingly important.

10.4 REFERENCES

Aigouy, B.M. (2013). ScientiFig: a tool to build publication-ready scientific figures. *Nature Methods*, 10, 1048 doi:10.1038/nmeth.2692.

Allan, C. et al. (2012). OMERO: a flexible, model-driven data management system for experimental biology. *Nature Methods*, 28; 9: 245–253. doi: 10.1038/nmeth.1896.

AwareSystems. (n.d.). Retrieved from http://www.awaresystems.be/imaging/tiff.html

Blatt, M. (2013). Manipulation and Misconduct in the Handling of Image Data. *The Plant Cell and Plant Physiology*, 25 no. 9-3147-3148 http://dx.doi: 10.1105/tpc.113.250980.

Goldberg, I.G. et al. (2005). The Open Microscopy Environment (OME) Data Model and XML file: open tools for informatics and quantitative analysis in biological imaging. *Genome Biology*, 6:R47 DOI: 10.1186/gb-2005-6-5-r47.

Linkert, M. et al. (2010). Metadata matters: access to image data in the real world. *Journal of Cell Biology*, 189, 5777–782 doi: 10.1083/jcb.201004104.

North A.J. (2006). Seeing is believing? A beginners' guide to practical pitfalls in image acquisition. *Journal of Cell Biology*, 172:9–18. doi:10.1083/jcb.200507103.

Williams, E. et al. (2017). Image Data Resource: a bioimage data integration and publication platform. *Nature Methods*, 14, 775–781 doi:10.1038/nmeth.4326.

Yamada, R. (2004). What's in a picture? The temptation of image manipulation. *Journal of Cell Biology*, 166 no. 1–11-15 doi: 10.1083/jcb.200406019.

11

Epilogue: A Framework for Bioimage Analysis

Kota Miura[1,2,4] and Sébastien Tosi[3,4]

[1] *Nikon Imaging Center, Bioquant, University of Heidelberg, Germany*
[2] *National Institute of Basic Biology, Okazaki, Japan*
[3] *Advanced Digital Microscopy Core Facility (ADMCF), Institute for Research in Biomedicine (IRB Barcelona). The Barcelona Institute of Science and Technology, Barcelona, Spain*
[4] *Network of European Bioimage Analysts (NEUBIAS)*

In this final chapter we aim to guide you towards the customised assembly of image processing components in practical workflows for analysing biological image data. We will first introduce the concepts of workflows and components and then point you to accessible resources, which we believe are helpful for the designing of the image analysis strategy and its implementation. A detailed view into how to write scripts or to implement image processing algorithms is covered in Our book Bioimage Analysis (see below), so here we will draw the outlines of this design process and its fundamental goals and pitfalls. For more practical information beyond the scope of this book, please refer to our freely downloadable book from the publisher's website (Miura et al. 2016) and to the NEUBIAS website (www.neubias.org).

11.1 WORKFLOWS FOR BIOIMAGE ANALYSIS

In the vast majority of research projects involving bioimage analysis, the goal is either to quantify some target biological phenomena or to visualise multidimensional structures for a better understanding of the underlying biological systems. (We use the terms 'image analysis' and 'bioimage analysis' distinctively, the former primarily aims to mimic human visual recognition, while the latter aims to perform quantitative measurements on biological systems – Miura and Tosi 2016.) Visualisation is a reduction of multidimensional data to lower dimensional data, typically 2D, to display it in a comprehensive way for our brain. Image measurements provide quantitative support for the results and hypothesis presented in scientific articles. This roots back from physiological measurements of microscope images (M Dougall 1897, Newman 1879) and from biochemical measurements (Schnell 2003, Cornish-Bowden 2013) started in 19th century. What is new, is that we are now combining the acquisition of multidimensional digital images from modern automated instrumentation with computation to measure the complexity of biological systems per se, and analyse it in a quantitative manner. Instead of sketching cells and measuring protein density in beakers as in the early days of biological research, we now combine various image processing algorithms, which are typically designed to enhance and delineate some specific spatiotemporal structures in the images, and measure some of their properties (e.g. average intensity) in a specific imaging channel, their geometry, location, frequency of occurrence and possibly the dynamics of all these measurements. Modern computers enable the automation of these tasks and can transform a huge amount of data, both in raw size and number, into meaningful statistical measurements.

To efficiently employ this computational power, we need to combine image processing and data analysis algorithms in a logical and structured manner to pipeline the processing from the original multidimensional data into valuable numerical measurements. We will refer to this pipeline as a 'workflow' and each of its constitutive algorithms as 'components'. The successful construction of a good workflow is the key to successful bioimage analysis and can also provide an accurate and readable protocol to reproduce the analysis undertaken by a research group. In the following sections we deepen this view on workflows and components in the context of bioimage analysis.

11.1.1 Components

A component is the implementation of a certain image processing, analysis or data analysis algorithm. A component usually does not solve a whole analysis problem on its own, but implements an elementary operation.

Table 11.1 Types of components.

Type	Input	Output	Examples
Image processing components	Image	Image	Gaussian blur filter Otsu thresholding
Image analysis components	Image (and sometimes annotations)	Annotations	Intensity local maxima detection Seeded watershed Connected components analysis
Data analysis components	Annotations (and sometimes image)	Numerical data (flat table or structured data).	Intensity statistics Object clustering labels Cell lineage tree Positions Fitted parameters

Essentially three types of components – image processing, image analysis and data analysis components – are used to build bioimage analysis workflows (Table 11.1).

Note that the word 'annotation' in this article is used in its most generic meaning, that is either manually or automatically generated geometrical models. These models can for instance be defined as what normally is called regions of interest (ROIs – closed geometrical contours or surfaces), wires representing the centreline of a network, trajectories and a list of coordinates to locate markers, and this encompasses binary masks and segmented images.

In practice, the concrete implementation of these components are accessible as:

- a menu item in an image processing software
- a plugin/module/add-ons of some image processing software
- a class or method in an image processing library
- a standalone executable called from the command line.

The majority of these components are packaged within standalone software with dedicated graphical user interface (GUI), or in a library to be accessed by other computer software.

Most software and libraries only provide a flat list of components without any pre-assembled workflows, and the user is left alone to finish the LEGO without a master plan. Image analysis beginners often start blindly choosing various menu items, testing around arbitrary components and looking for some encouraging results that loosely match their expectations. There is a strong reason for this wandering: examples, methods and guidelines for workflow designing are missing.

11.1.2 *Workflows*

A bioimage analysis workflow is a set of *components* assembled in some specific order to process bioimages to estimate some numerical parameters relevant to the biological systems under study. The wiring of components in a workflow can be as simple as a linear path or as complex as many branchings and mergings, involving loops, condition testing and user interactions. Workflows are usually assembled from components coming from a single software package, but sometimes are taken and gathered from multiple software packages. The workflow can be constructed in various forms, such as:

- a script of a programming environment (e.g. MATLAB, R, Python, Shell) or a platform supporting scripting (e.g. ImageJ, Icy) that calls components of the software by following a user defined sequence
- a flowchart diagram in a graphical programming interface, made of components as boxes and arrows representing the data flow between components (e.g. Labview, Icy protocol, KNIME)
- a GUI allowing the interactive stacking and configuration of components to build a pipeline (e.g. CellProfiler, ilastik)
- detailed step-by-step instructions in text on how to manually operate a sequence of operations to process the data using one or several software packages (methods section in scientific article).

A well-written workflow allows other researchers to reproduce the same results from the same data. Completely reproducible workflows such as scripts, graphical programming files and pipeline files are recommended to be published as scientific documentation, together with results. Sadly, many published papers only mention the name of the software used for the analysis in the methods section and give no clear workflow. Such poor and unscientific practice should absolutely be avoided.

ImageJ and MATLAB come only with a flat list of components, but since they are popular in the field of bioimage analysis many workflow scripts are available from their huge user communities. Although they are often tailored specifically to the conditions and needs of certain project, a well-written and documented workflow can be easily modified to address a slightly different problem, or at least partly reused as a code base or a source of inspiration. Some software, such as CellProfiler, Volocity (currently unsupported), MetaExpress (Molecular Devices) and

NIS Elements (Nikon Instruments) readily provide some popular types of workflows and help the users to tweak or customise them instead of starting from scratch.

11.1.3 Types of Workflows

Although each workflow is tailored and uniquely associated to a specific project, there are broad classes of workflow. Many bioimage analysis workflows follow a core sequence of steps: loading image dataset, pre-processing, annotation (e.g. segmentation), post-processing, measurements and plotting or visualisation of results. This sequence can be more complex, but for now let's stick to this simple structure, and study some workflow variants around it.

The simplest workflow deals with image data holding only a single imaging channel. This could be a bright-field image, or a fluorescence image with one molecular species labelled. The goal of the analysis is to measure some feature of a given type of target object in the image data, e.g. measurement of the distribution of cell nucleus areas. The workflow for this task is linear, since we only need to pre-process images to simplify their segmentation, perform the segmentation to delineate the object and finally measure the target feature. We call this 'workflow type 0' (Figure 11.1, right column top). Since this type of workflow is straightforward, it tends to be packaged into a single tool or plugin with a set of adjustable parameters of its components. One example of this is particle tracking which typically starts with particle position detection, and proceeds to a thick layer of data analysis component involving linking of those positions along time to recover trajectories (Tinevez 2016, Sbalzarini 2005). Note that the simplicity of a workflow does not necessarily correlate with the complexity of each of its constitutive components.

A slight complexity is added when our target feature is intensity, e.g. a fluorescence signal. We then need to have two different paths in the workflow. Firstly, we need a path that annotates the target object or structure. This is required for determining the regions for measuring the intensity. Secondly, we need another path for measuring the intensity. Processing of the image in this path should be carefully planned but some components might improve the quantitative measurement in some situations (e.g. denoising, deconvolution). Note that nonlinear operations should either be avoided or at least well calibrated before being used in this path. Finally, we can measure the signal intensity in the second path inside the measurement regions (annotations) obtained in the first path. This leads to a workflow architecture with two converging

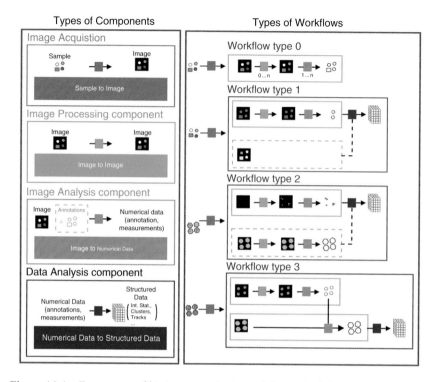

Figure 11.1 Four types of bioimage analysis workflow. The left column shows the schematic definition of component types (also refer to Table 11.1). The right column shows four type of workflows as flowcharts: Type 0, the simplest workflow segmenting a single type of object from a single channel; Type 1, single- or dual-channel dual-branch workflow annotating a specific object type and then measuring intensity inside these objects in the raw images; Type 2, dual-channel dual-branch co-measuring two object types; Type 3, dual-channel dual-branch detecting seeds in one channel to segment objects in the other channel. Original design: Romain Guillet.

paths that we will call 'workflow type 1' (Figure 11.1). Hence, for this type of workflow, we can use any processing components in the first path as long as it leads to a successful delineation of the target structures, but in the second path we should be careful enough not to skew or lose the quantitative information in the original image. A classic example of this type of workflow can be found in Hirschberg et al. (1998) and more recently Boni et al. (2015). The latter paper reports measurements of intensity dynamics of an inner nuclear membrane at nuclear envelope and endoplasmic reticulum, with complete workflow codes available in supplementary information.

Dual-path workflows are also prevalent when detection and measurements are to be performed on different structures. In many research projects, we aim to measure some feature of a structure A, that is somehow encapsulated in structure B. The feature to be measured could be very varied, such as how many A are in each B, or localisation of structures A within the space bounded by structures B. In such cases, we typically capture two image channels, one for each structure, and then separately design an annotation path for each. Upon successful delineation of the objects, the annotations from the two paths are combined to lead to the co-measurement. We call such workflows 'workflow type 2' (Figure 11.1). Example applications can be fluorescence in-situ hybridisation (FISH) images, to count the number of FISH signals per nucleus, or to analyse gene loci distance from the nuclear envelope to evaluate the gene expression activity and the location of that gene within the nucleus (Vaquerisas 2010).

Two-channel image data is also used in other ways, such as to simplify the segmentation of a single type of structure in one of the two channels. A typical workflow of this type uses structure A encapsulated in structure B as a seed (marker) to segment B. This is the case, for instance, when a cell nucleus signal (A) in the first channel is used to segment a single cell signal (B) in the second channel, and then this segmented cell boundary is used for further measurement of certain features such as cell shape and movements. We call this architecture 'workflow type 3' (Figure 11.1). This type of workflow is widely used for segmenting cell boundaries in multicellular tissues (Paul-Gilloteaux 2016) and for densely populated cells in cell culture (Jones 2006).

Importantly, the measurement targets are sometimes not accessible by direct image measurements but can only be inferred by building a system model. The final measurements are parameters of this model and are typically estimated by statistical optimisation or fitting of predicted values to the direct measurements. This type of measurement includes a thick layer of data analysis components in the final steps of the workflow. Example workflows involving such data analysis components are FRAP analysis (Mueller 2010), tissue tension estimation from retraction speed profile (Colombelli 2009), force estimation from velocity fields estimated by PIV (Hernández-Vega 2016) and as already mentioned in type 0, object tracking.

Finally, we should note that there are many more workflows not fitting into these four types but they already capture a large majority of practical bioimage analysis workflows.

11.1.4 *Types of Component*

The components constituting workflows are generally written as optimised and compact code in low-level languages (typically C, C++ or Java). Each component is the implementation of a specific algorithm and, even though that might be difficult, it is always important to have some understanding of what each component is doing. Rather than listing all existing components, we have grouped components by their function and mapped them in an archetypal image analysis workflow for an overview (Figure 11.2). These component groups are wired so as to integrate all types of workflows detailed in the previous section.

In general, the images are first accessed from files stored on the computer (data reader components) and, when applicable, recombined to larger images that are not necessarily limited to the field of view of the instrument (reconstruction components). Restoration components compensate for acquisition artefacts such as image noise, uneven illumination,

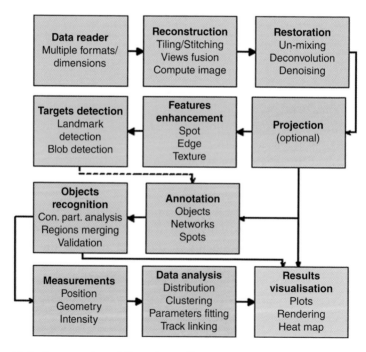

Figure 11.2 Integrated workflow chart of bioimage analysis. Each box is a component type, and the colours correspond to the three conceptual types defined in Table 11.1. Pink and grey: image processing components; Blue and orange: image analysis components; Blue-grey and purple: data analysis components.

channel crosstalk and the blurring induced by diffraction-limited instruments. A critical decision is then whether the analysis is to be performed on the raw data or on projected data (e.g. 3D z-stack z projection). In practice, this decision is driven by the complexity of the sample and the task to achieve. Then there might be steps of feature enhancement to increase the contrast of the structures of interest such as spots, filaments and contours against other structures and background (features enhancement components). This is sometimes followed by targets detection components, automatically finding a marker, or seeds, inside each object of interest. Next, the objects are delineated to create annotations (annotation components), for instance in the form of a segmentation mask in which pixel values are binarised, e.g. black and white, to mark background and foreground regions (object recognition components). Finally, the annotations can be refined by merging connected foreground regions, discarding objects not fulfilling the expected geometry of the structures of interest, or splitting them in parts that better fulfil the expected criteria. The last steps typically involve object measurements (measurements component, e.g. morphometry, intensity), some statistical analysis of these measurements (data analysis components), and the output of the image analysis results in various visualisation formats such as distributions, plots, object outlines and heat maps (results visualisation components).

We can never stress enough that the validation of the analysis quality (and possibly error correction, a.k.a. curation) is essential. Since visual inspection of the results of the whole dataset is often prohibitive, it should at least be performed on a fraction of the data and by including some representative regions of the whole data. Building efficient and user-friendly tools for the exploration, validation and correction of automatic image analysis is often one of the most complex parts of a practical workflow, especially for large multidimensional dataset for which interactive visualisation is already a challenge per se.

11.2 RESOURCES FOR DESIGNING WORKFLOWS AND SUPPORTING BIOIMAGE ANALYSIS

In the previous section we only scratched the surface of how components can be assembled to lead to bioimage analysis workflows. To construct a workflow tailored to a specific project, it is important to first pick or design an outline of the workflow. Then each component should be carefully selected to achieve the best results for that step and finally assembled. To facilitate the design of workflows, there is an initiative to build a community of bioimage analysis. Through this community we are actively preparing textbooks, organising courses and constructing a database of components and workflows.

In the following sections, we briefly introduce each of these activities, in the hope that it will be a valuable resource for the readers to design their own analysis workflows. At the same time, we hope that this will motivate your engagement in the activity of our community in the future!

11.2.1 *A Brief History*

Bioimaging modalities have become increasingly diverse and complex in recent years, from the inception of open-source Bioimage analysis tools in the 1970s such as NIH Image, by Wayne Rasband, which then became ImageJ (Schneider 2012) to the vast range of modalities available to researchers currently, detailed in Chapter 1. The power of imaging is constantly increasing, with the emergence of new instruments and methods discussed elsewhere in this book. Algorithm developers have been working hard to provide increasingly better components that are useful bricks of powerful image analysis workflows. These efforts are providing practical solutions for some image analysis tasks. However, the increase in the sheer number of components, platforms and dedicated software can make the quest for the optimal solution confusing, especially to life scientists with little or no background in the field.

Overall, there is a lack of guidelines to efficiently choose and assemble existing components. At the same time, the uniqueness of each research project produces a wide diversity in the image analysis tasks, which calls for the customisation of readily available workflows. For this reason, experts working on constructing bioimage analysis workflows are starting to emerge in the life science research community, to collaborate with life scientists to tackle this difficulty.

As these experts in bioimage analysis are part of a new profession within life science, many of them do not have strong ties with their peers. For this reason, with a small group of such experts, we started to plan a series of bioimage analysis courses together supported by the course and conference office of the European Molecular Biology Laboratory. Through rounds of discussion during the first course preparation in 2013, we realised that many ingredients that could facilitate bioimage analysis activity in life science were missing.

11.2.2 *A Network for Bioimage Analysis*

Experts in bioimage analysis have often been considered to only be algorithm developers. While analysis experts concentrate on assembling multiple components to design workflows for specific life science research projects, the work of developers is rather focused on the engineering of sophisticated and efficient image processing algorithms towards scientific publication. That is, they mostly design components. These algorithm developers have a long tradition dating back from the field of computational science and have a firm academic root and tie within their own community of colleagues.

On the other hand, these new bioimage analysis experts have had difficulty in positioning their expertise in between life science and computational science, and actually do not fully belong to either of these. Their role is interdisciplinary and they are called to act as bridges between these disciplines and the community of bioimaging, where most are microscope application specialists. This is reflected in their huge heterogeneity in position titles, contract types, backgrounds and job descriptions. For this reason, to foster a stronger identity we started to call ourselves 'bioimage analysts', and also started to network with others working on designing bioimage workflows. In March 2016, this network became an EU funded programme – the Network of European Bioimage Analysts (NEUBIAS: http://www. cost.eu/COST_Actions/ca/CA15124), and it is now actively promoting the identity of bioimage analysts and facilitating information exchange among professionals.

11.2.3 *Additional Textbooks*

Life scientists who go astray while trying to analyse their own image data often become confused or frustrated as they try to find help in digital image processing textbooks. The main reason is that most conventional image processing textbooks are centred on mathematical details and insights into algorithms of components and not on the designing of bioimage analysis workflows. As an example, image filters are inevitable components, but their algorithmic details alone do not provide any direct clues as to how they could be used in a workflow. For these reasons, the network of bioimage analysts started to compile a textbook dedicated to bioimage analysis, which is centred on the practical implementation of bioimage analysis workflows and scripting (Miura et al. 2016). At the moment, we are further working in the same direction to collect, organise and publish more workflows that could become usable templates for customising workflows for original research projects, and these will be compiled as other editions of similar textbooks. The reader who is interested in developing their skills in scripting and coding is recommended to further the knowledge from this book, *Bioimage Data Analysis* (K. Miura, Ed. Weinheim: Wiley-VCH, available from http://www.imaging-git.com/applications/bioimage-data-analysis-0)

11.2.4 *Training Schools*

In the same way that textbooks have been missing, practical bioimage analysis courses centred on the needs of the life science community are very rare. While computational analysis of image data is becoming more and more widespread in life science research, the university curriculum is generally lacking the teaching of the fundamentals required for the quantitative measurement of image data. As evidence of this, a survey conducted by the network of bioimage analysts in 2015 showed that ~68% of responding researchers from the European bioimaging community considered support and training for bioimage analysis to be missing (Figure 11.3).

In order to fulfil the need, we started a bioimage analysis course with a small number of bioimage analyst colleagues in 2013. It has now evolved into one of the central activity of NEUBIAS, which organises three different courses for (1) life scientists and (2) imaging facility staff who are experienced in acquiring image data but missing good practice and tools to build their own workflow, and for (3) active bioimage analysts wishing to build up their skills. All these courses focus on the design of workflows. Courses are also an excellent opportunity to share workflows among fellow bioimage analysts who are instructors of the course. At the same time, as bioimage analysts are mostly working alone or in small groups of institutes, these courses have become an open gateway to the community of bioimage analysis through collaboration in designing and running the course.

11.2.5 *Database of Components and Workflows*

Another important action taken by NEUBIAS is to set up a web platform with searchable knowledge base for assisting the construction of workflows. In this platform, components and workflows will be cross-linked and tagged by terms coming from both image processing and life science

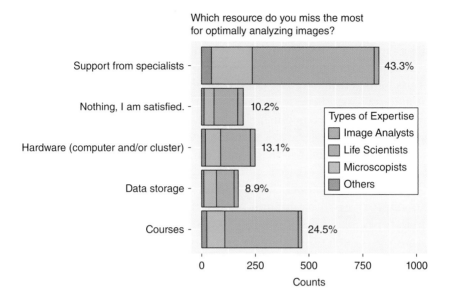

Figure 11.3   Results of a survey in 2015. To the question 'Which resource do you miss the most for optimally analysing images?', 67.8% of respondents, mostly from the bioimaging community, chose either support from specialists or courses (n = 1904). The colors are for different expertise identified by each of the responders by themselves.

communities. As these communities use their own specialist terms, onto-logical glossaries will bridge image analysis components, workflows and questions to be addressed in life science studies, and help to fill the vocabulary gap. We call this platform a 'Web tool', and its pilot version is currently administered and edited by NEUBIAS members (http://biii. info and its developer version http://biii.eu). In the future it will be fed with more content coming from crowdsourcing. In practice, life scientists are often limited by their own knowledge of technical keywords when trying to search in image-processing-related forums for components or workflows to achieve some image analysis task. The Web tool provides a practical solution to this problem that is typically faced in growing interdisciplinary fields. Moreover, publicly available workflows are currently spread across websites, method sections in research papers and code repositories, and are very often in a poorly usable form. Curation of these resources in the Web tool will bring some standardisation to the style of their documentation and provide some explicit specification to their interface, enabling a more efficient designing and recycling of bioimage analysis workflows.

11.2.6 *Benchmarking Platform*

When building a workflow and selecting its components, there are often some daunting questions: How close to optimality am I? Are the results I am getting already 'good enough' for my study? In the field of computer vision there is a current tradition of objectively assessing the performance of components, which is called benchmarking. Performance can be as generic as execution speed, memory requirements or limited to a certain problem, such as specificity and sensitivity of object detection. It is often summarised by numerical values that are called metrics; developers are as if they are at open war to incrementally improve performance and publish new components together with comparative benchmark results. To ease this process and to increase its relevance, several challenges were organised during International Symposium on Biomedical Imaging (ISBI, https://grand-challenge.org/all_challenges/, Kozubek 2016). In particular, we can mention deconvolution challenges (2013 and 2014), single particle tracking challenge (2012) and single molecule localisation challenges (2014 and 2016). Looking forwards, NEUBIAS plans a repository of standard image datasets representing bioimage analysis problems. At the same time, it will identify the most relevant metrics for each class of problem and provide the benchmark of existing workflows. Ease of deployment, usability and robustness of benchmark results to various image qualities will also be taken into account. This newly created benchmarking web module will be integrated tightly to the sample image and component/workflow databases so that after identifying a candidate workflow it will be simple to test it on reference images and compare the results to the results of other workflows.

11.3 CONCLUSION

In this epilogue we have tried to overview the current status of how bioimage analysts are working to provide solutions to the life science community. It might have been a bit far off from practical implementation, but we hope that this appetiser will help you find your way and trigger your curiosity to further explore this fascinating field. To achieve an outstanding scientific goal, both originality of the questions and well thought out approaches are required. In this respect, image analysis, which is among the very last steps of imaging-based research projects, cannot be accomplished as an off-the-shelf routine procedure. We then expect that image analysis in every project requires a certain degree of optimisation, customisation or full-scratch workflow design to provide relevant answers. Hopefully, this epilogue and the resources we have pointed to will help you to become more efficient in designing your own bioimage analysis workflows, for a smarter struggle towards questioning biological systems.

11.4 REFERENCES

Boni, A., Politi, A. Z., Strnad, P., Xiang, W., Hossain, M. J. and Ellenberg, J. (2015). Live imaging and modeling of inner nuclear membrane targeting reveals its molecular requirements in mammalian cells. *The Journal of Cell Biology*, 209(5), 705–720.

Colombelli, J., Besser, A., Kress, H., Reynaud, E.G., Girard, P., Caussinus, E., et al. (2009). Mechanosensing in actin stress fibers revealed by a close correlation between force and protein localisation. *Journal of Cell Science*, 122(Pt 10), 1665–79.

Cornish-Bowden, A. (2013). The origins of enzyme kinetics. *FEBS Letters*, 587(17), 2725–30.

Hernández-Vega, A., Marsal, M., Pouille, P.-A., Tosi, S., Colombelli, J., Luque, T. et al. (2016). Polarised cortical tension drives zebrafish epiboly movements. *The EMBO Journal*.

Hirschberg, K., Miller, C.M., Ellenberg, J., Presley, J.F., Siggia, E.D., Phair, R.D. and Lippincott-Schwartz, J. (1998). Kinetic analysis of secretory protein traffic and characterisation of golgi to plasma membrane transport intermediates in living cells. *The Journal of Cell Biology*, 143(6), 1485–503.

Jones, T.R., Carpenter, A.E., Sabatini, D.M. and Golland, P. (2006). Methods for high-content, high-throughput image-based cell screening. In D. Metaxas, R. Whitaker, J. Rittcher and T. Sebastian (Eds), *Proceedings of the First MICCAI Workshop on Microscopic Image Analysis with Applications in Biology* (pp. 65–72). Copenhagen, Denmark.

Kozubek, M. (2016). Challenges and Benchmarks in Bioimage Analysis. *Advances in Anatomy, Embryology, and Cell Biology*, 219, 231–62.

M'Dougall, W. (1897). The Structure of Cross-Striated Muscle, and a Suggestion as to the Nature of its Contraction. *Journal of Anatomy and Physiology*, 31(Pt 4), 539–85.

Miura, K., Tosi, S., Möhl, C., Zhang, C., Paul-Gilloteaux, P., Schultz, U. et al. (2016). *Bioimage Data Analysis*. (K. Miura, Ed.). Weinheim: Wiley-VCH. Retrieved from http://www.imaging-git.com/applications/bioimage-data-analysis-0

Miura, K., and S. Tosi. (2016). Introduction. *In* Bioimage Data Analysis. K. Miura, editor. Wiley-VCH, Weinheim. 1–3.

Mueller, F., Mazza, D., Stasevich, T.J. and McNally, J.G. (2010). FRAP and kinetic modeling in the analysis of nuclear protein dynamics: what do we really know? *Current Opinion in Cell Biology*, 22(3), 403–11.

Newman, D. (1879). New Theory of Contraction of Striated Muscle, and Demonstration of the Composition of the Broad Dark Bands. *Journal of Anatomy and Physiology*, 13(Pt 4), 549–76.

Paul-Gilloteaux, P. and Tosi, S. (2016). Quantitative Evaluation of Multicellular Movements in Drosophila Embryo. In K. Miura (Ed.), *Biomage Data Analysis* (pp. 170–197). Wiley-VCH.

Sage D., Kirshner H., Pengo T., Stuurman N., Min J., Manley S. and Unser M. (2015). Quantitative evaluation of software packages for single-molecule localisation microscopy. *Nature Methods* 12, 717–724 doi:10.1038/nmeth.3442

Sbalzarini, I.F., Mezzacasa, A., Helenius, A. and Koumoutsakos, P. (2005). Effects of organelle shape on fluorescence recovery after photobleaching. *Biophysical Journal*, 89(3), 1482–1492.

Schnell, S. and Maini, P.K. (2003). A century of enzyme kinetics. Reliability of the KM and vmax estimates. *Comments on Theoretical Biology*, 8(2), 169–187.

Schneider, C.A, Rasband, W.S, Eliceiri, K.W. (2012). NIH Image to ImageJ: 25 years of image analysis. *Nature Methods* 9, 671–675 Doi: 10.1038/nmeth.2089

Tinevez, J.-Y., Perry, N., Schindelin, J., Hoopes, G.M., Reynolds, G.D., Laplantine, E. et al. (2016). TrackMate: An open and extensible platform for single-particle tracking. *Methods*.

Vaquerisas, J.M., Suyama, R., Kind, J., Miura, K., Luscombe, N.M. and Akhtar, A. (2010). Nuclear pore proteins Nup153 and megator define transcriptionally active regions in the Drosophila genome. *PLoS Genetics*, 6(2), e1000846.

Index

Page numbers in *italics* refer to figures; those in **bold type**, to tables.

Standard and Super-Resolution Bioimaging Data Analysis: A Primer, First Edition.
Edited by Ann Wheeler and Ricardo Henriques.
© 2018 John Wiley & Sons Ltd. Published 2018 by John Wiley & Sons Ltd.